# Modeling and Simulation Techniques for Improved Business Processes

Maryam Ebrahimi
*Azad University, Iran*

A volume in the Advances in Business Information Systems and Analytics (ABISA) Book Series

IGI Global
DISSEMINATOR OF KNOWLEDGE

Published in the United States of America by
IGI Global
Business Science Reference (an imprint of IGI Global)
701 E. Chocolate Avenue
Hershey PA, USA 17033
Tel: 717-533-8845
Fax: 717-533-8661
E-mail: cust@igi-global.com
Web site: http://www.igi-global.com



Library of Congress Cataloging-in-Publication Data

Names: Ebrahimi, Maryam, 1980- editor.
Title: Modeling and simulation techniques for improved business processes / Maryam Ebrahimi, editor.
Description: Hershey : Business Science Reference, [2017]
Identifiers: LCCN 2017015920| ISBN 9781522532262 (hardcover) | ISBN 9781522532279 (ebook)
Subjects: LCSH: Management--Simulation methods. | System theory. | Operations research.
Classification: LCC HD30.23 .M594 2017 | DDC 658.4/032--dc23 LC record available at https://lccn.loc.gov/2017015920

British Cataloguing in Publication Data
A Cataloguing in Publication record for this book is available from the British Library.

# Advances in Business Information Systems and Analytics (ABISA) Book Series

ISSN:2327-3275
EISSN:2327-3283

Editor-in-Chief: Madjid Tavana, La Salle University, USA

**MISSION**

The successful development and management of information systems and business analytics is crucial to the success of an organization. New technological developments and methods for data analysis have allowed organizations to not only improve their processes and allow for greater productivity, but have also provided businesses with a venue through which to cut costs, plan for the future, and maintain competitive advantage in the information age.

The **Advances in Business Information Systems and Analytics (ABISA) Book Series** aims to present diverse and timely research in the development, deployment, and management of business information systems and business analytics for continued organizational development and improved business value.

**COVERAGE**

- Statistics
- Strategic Information Systems
- Forecasting
- Business Information Security
- Data Governance
- Geo-BIS
- Management Information Systems
- Data Management
- Information Logistics
- Business Models

IGI Global is currently accepting manuscripts for publication within this series. To submit a proposal for a volume in this series, please contact our Acquisition Editors at Acquisitions@igi-global.com or visit: http://www.igi-global.com/publish/.

The Advances in Business Information Systems and Analytics (ABISA) Book Series (ISSN 2327-3275) is published by IGI Global, 701 E. Chocolate Avenue, Hershey, PA 17033-1240, USA, www.igi-global.com. This series is composed of titles available for purchase individually; each title is edited to be contextually exclusive from any other title within the series. For pricing and ordering information please visit http://www.igi-global.com/book-series/advances-business-information-systems-analytics/37155. Postmaster: Send all address changes to above address. 

## Titles in this Series

*For a list of additional titles in this series, please visit:*
*https://www.igi-global.com/book-series/advances-business-information-systems-analytics/37155*

***Cloud Computing Technologies for Green Entrprises***
Kashif Munir (University of Hafr Al-Batin, Saudi Arabia)
Business Science Reference • ©2018 • 424pp • H/C (ISBN: 9781522530381) • US $215.00

***Smart Technology Applications in Business Environments***
Tomayess Issa (Curtin University, Australia) Piet Kommers (University of Twente, The Netherlands) Theodora Issa (Curtin University, Australia) Pedro Isaías (Portuguese Open University, Portugal) and Touma B. Issa (Murdoch University, Australia)
Business Science Reference • ©2017 • 429pp • H/C (ISBN: 9781522524922) • US $210.00

***Advanced Fashion Technology and Operations Management***
Alessandra Vecchi (London College of Fashion, University of London Arts, UK)
Business Science Reference • ©2017 • 443pp • H/C (ISBN: 9781522518655) • US $200.00

***Maximizing Business Performance and Efficiency Through Intelligent Systems***
Om Prakash Rishi (University of Kota, India) and Anukrati Sharma (University of Kota, India)
Business Science Reference • ©2017 • 269pp • H/C (ISBN: 9781522522348) • US $175.00

***Maximizing Information System Availability Through Bayesian Belief Network Approaches...***
Semir Ibrahimović (Bosna Bank International, Bosnia and Herzegovina) Lejla Turulja (University of Sarajevo, Bosnia and Herzegovina) and Nijaz Bajgorić (University of Sarajevo, Bosnia and Herzegovina)
Information Science Reference • ©2017 • 180pp • H/C (ISBN: 9781522522683) • US $140.00

***Handbook of Research on Advanced Data Mining Techniques and Applications...***
Shrawan Kumar Trivedi (BML Munjal University, India) Shubhamoy Dey (Indian Institute of Management Indore, India) Anil Kumar (BML Munjal University, India) and Tapan Kumar Panda (Jindal Global Business School, India)
Business Science Reference • ©2017 • 438pp • H/C (ISBN: 9781522520313) • US $260.00

*For an enitre list of titles in this series, please visit:*
*https://www.igi-global.com/book-series/advances-business-information-systems-analytics/37155*

IGI Global
DISSEMINATOR OF KNOWLEDGE

701 East Chocolate Avenue, Hershey, PA 17033, USA
Tel: 717-533-8845 x100 • Fax: 717-533-8661
E-Mail: cust@igi-global.com • www.igi-global.com

# Table of Contents

# Detailed Table of Contents



In response to uncertainty of prospective technologies and how they might fit market demand, firms tend to establish R&D alliances. In this chapter the effect over time of continuation of underperforming R&D alliances on innovation performance during the pre-market stage is investigated. This stage is characterized by non-linearity, as expected outcomes and market demands are uncertain. Literature suggests that computational modeling in particular agent-based modeling can be used to investigate such non-linear processes. Agent based modeling starts with simple behavioral rules that develop into emergent system-level behaviors, and in that way controlled system level experiments are used to identify in an inductive way causal mechanisms that drive the system development. In this chapter's simulation model, an agent decides to continue its R&D alliance based on its strategic and cooperation objectives. After evaluating if the strategic goals is met, firms can decide about the extent to which to continue the R&D alliances if the strategic goal is not met. This is called persistency. The model is aimed to explain developmental paths and patterns of the co-evolution of alliances and technology. Despite suggestions to investigate non-linear processes in the pre-market phase by using an agent-based model, agent-based models so far do not focus on the impact of alliance continuation on innovation performance over the path of technology development. In previous research these paths mainly have been investigated in case and cross sectional studies but not in an agent based model. A base-line model is developed and the extent to which it reflects reality is analyzed in order to improve the model's performance.

The main purpose of this study is to describe and analyze an agent from a distributed multi-agent based system (ABS) according to the BDI architecture. This agent is capable of autonomous action to propose general technology strategies (TSs) in renewable energy SMEs based on a set of rules and interacts with a core agent in multi ABS. The recognition of internal strengths and weaknesses as well as external opportunities and threats takes place on the basis of technological SWOT-analysis. Proposed TSs are categorized into four types: aggressive strategy, turnaround oriented strategy, diversification strategy, and defensive strategy. Agent architecture in terms of three abstraction layers called psychological, theoretical, and implementation is explained. And after system validation by experts, some program codes and output results of this agent are presented. This system provides information to facilitate the TS planning process to be carried out effectively.



Localizing and creating technical knowledge in the developing countries to make them able to compete in the international arena will not be an easy task to undertake as it involves complexities and uncertainties. In this regard, methods of modeling and simulation help understanding the reality and thereby delineating the future. This paper discusses using system dynamics methodology in the Petrochemical Research and Technology Company (Iran). After determining causal loop diagrams, a dynamic model, and validation test, in order to improve the system, related scenarios to the research are proposed. The results show that research financing is a major obstacle in the course of growth and development; accordingly, some scenarios are recommended to increase technical knowledge through raising the initiation and completion rates of research projects. The paper closes with a description of the suggested scenarios.



The analysis of human resources function and its contribution to obtain performance dates back to the 1920s. Now, the HR is an equal partner on the board of the companies, having a strategic role in obtaining performance, thus we try to show

such as the implementation of policies, making quota decisions, identification and problem solving, response to uncertainty, among others. Regarding the resources dimensioning in hospitals, the Brazilian scenario is limited to legislative instruments that assume a prior and added sizing. This chapter uses a discrete event simulation tool to set the amount of operation rooms needed for patient care in an emergency department, so that emergency patients have guaranteed compliance, minimizing the cancellation of elective surgeries because of this type of demand. As a result, it was found that the minimum amount established by normative instruments was not appropriate to the specific requirements of the organization.



In its reality, the world is more complex than we expect it to be predicated and controlled; so, by minor view to issues and suppose many phenomena to be independent, it would be impossible to understand their totality and complexity. Utilizing appropriate scientific tools can provide us with useful information to manage decision making to improve methods to carry out operations and apply resources; one of these scientific tools is system dynamics approach. Therefore, based on stages in the methodology of system dynamics and by taking experts comments, in the present study, causal relations between influential factors in improvement of customs service are identified. So, by knowing these causal relations, one can allocate resources and prioritize activities leading to influence on other factors following with the best results. Finally, after validation and having the confirmation of the experts, the researcher has presented a causal model to improve customs services.



In this chapter, based on systems dynamics approach, efforts have been made to provide a model for the establishment and development of telework in public organizations of the Iran. The main goal of the study is to find a suitable and practical model for the establishment and successful development of telework in the public organizations of the country. Modeling for the establishment and development of telework for the first time in this research has been done using the systems dynamics approach. One of the main findings of this study is the extraction of the factors and main variables that are effective in the establishment and development of telework. Also, the effect and the relationship between the factors and operational variables of the model were revealed and the components of the causal loop were confirmed. Finally, the final model of the research was presented.

# Preface

## OVERVIEW

Current growth of businesses considering globalization and increased competition have led the organizations to reassess their businesses in order to make changes to them which is a key to the successful implementation of the essential adaptive changes. Additionally, the challenges of the recent environment have caused the need to a better understanding of sophisticated quantitative and analytical tools. Over the past decades, modeling and simulation techniques have become beneficial in designing and building complex systems. Computer simulation modeling is a discipline gaining popularity in both government and industry. Despite all the studies that have done in the area of socio-technical environment still the theory and in particular practice of modeling and simulation in business decision making, business process management, and technology and innovation management need more efforts.

This book aims to provide the theoretical foundation and practical practice of system dynamics, agent-based modeling, and discrete-event simulation in these research fields. System dynamics methodology provides an insight into the dynamic behavior of a system over time through dealing with interactions of various elements of a system in time and capturing the dynamic aspect by incorporating some concepts. Agent-based modeling and simulation is composed of autonomous, interacting agents and is a way to model the dynamics of complex systems and complex adaptive systems. Discrete event simulation portrays an event comprises a specific change in the system's state at a specific point in time and it codifies the behavior of a complex system as an ordered sequence of well-defined events.

Since these techniques are different in nature, they are often used for modeling different situations. Discrete event simulation is applied generally in systems which naturally involve queues of some sort, it tends to look at the smaller detail of a system (microscopic), and similar to agent-based modeling is stochastic in nature and therefore will give different results on different runs. System dynamics model is used more to model systems which naturally focus more on flows around networks than on the individual behaviour of entities, is deterministic in nature, so only needs to be run once, and tends to take a more overall perspective (macroscopic). Whilst system dynamics takes a top down approach, agent-based modeling is a very extreme example of a bottom up approach. Both agent-based modeling and discrete event simulation can use input distributions to model random behavior. Usually, agent-based modeling has been overlooked because of the tendency to use the other two methods. its models and systems are being made up of autonomous agents following a range of predetermined rules to achieve their objectives whilst having interactions with each other and their environment (Maidstone, 2012).

## APPLICATIONS

Globally, organizations are now seeking to find best ways to continually improve their products and services. Taking into consideration appropriate initiatives require usually large-scale changes to processes. The importance of customers and high intense competition have forced organizations to evaluate and improve customer-valued processes in a constant manner to be more efficient, effective, and flexible. Recently, most businesses recognize the prominance of adopting well-defined processes, and are now searching for carrying out those processes that create a competitive advantage. Thus, many organizations have started exploring methods to improve existing processes and develop new ones efficiently. Business process improvement has been developed to adapt to new conditions and competitive pressures. This concept has gained widespread acceptance as an effective approach to capture and communicate processes as a part of an organization's efforts to improve operational efficiency. Business process modeling and simulation are a useful tool to compare between "as-is" and "to-be" conditions with any methodology, in order to better analyzing the situation and showing the quantitative results after reducing the possible risks and drawing potential changes. Modeling and simulation techniques, such as system dynamics is helpful in such complex initiatives; they provide insights into feedback processes and lead to a better understanding of the dynamic behavior of the studied system. The application of system dynamics in this area concerns usually the achievement of the highest improvements in a system by graphical illustration of core business processes and the interactions within the organization. In addition, agent-based modeling and simulation is a valuable technique in simplifying the required efforts to understand the complexity of systems in many domains such as business process management. Discrete-event simulation can also be used to conduct business process analysis and it allows companies to test large number of data and explore hypothetical scenarios to define the best courses of action.

Business decision making is recognized as a complex dynamic multipurpose process and especially in enterprise with integrated structures and goals it needs some actions to reduce the possible risks of wrong decisions such as using modeling and simulation techniques. System dynamics is one way for challenging the increasing complexity of decision making in general and one of the best solution to avoid unintended consequences. Enterprise in agent-based modeling includes some decision-makers as agents with strategic, tactic and operational behavior. In addition, discrete-event simulation is applied in real-world decision-making.

This book also shows the possibility of studying innovation and technology management through applying modeling and simulation methods. System dynamics, agent-based modeling approaches are adopted to investigate the dynamics of research and development (R&D) projects, new product development projects, diffusion of innovation, and innovation and technological change. In this case, generally discrete event simulation is used with other modeling and simulation techniques as a hybrid modeling and simulation.

## TARGET AUDIENCE

This book is a collective resource with comprehensive knowledge in the field of organizational modeling and simulation in theory and practice. Computer simulation modeling can assist in the design, creation, and evaluation of complex systems. Designers, program managers, analysts, and engineers use computer simulation modeling to understand and evaluate 'what if' case scenarios. Target audiences are practitioners (business and system analysts, experts in modeling and simulation, managers of organizations), academics (researchers, lecturers and advanced students) and professional institutions (professional bodies of simulation) .

## ORGANIZATION OF THE BOOK

The book is organized into eight chapters. A brief description of each of the chapters follows:

Chapter 1, 'Innovation Success Over Time of Alliances With Different Strategic and Cooperation Objectives', written by Khrisna Ariyanto Manuhutu, Ariane von Raesfeld, and Peter Geurts, investigates the effect of continuation of underperforming R&D alliances on innovation performance over time during the pre-market stage by using agent-based modeling. The model explains developmental paths and patterns of the co-evolution of alliances and technology.

Chapter 2, 'BDI Approach to Build a Single Agent of a Distributed Multi-Agent System', written by Maryam Ebrahimi, describes an agent of a multi-agent system for the case of technology strategic planning by using BDI architecture. It focuses on application domains which is multi-agent decision systems in the area of strategic planning.

Chapter 3, 'Scenario Planning for Technical Knowledge Development', written by Maryam Ebrahimi, concerns technical knowledge development in an R&D organization and uses system dynamics methodology to illustrate the complexities of R&D processes and generate and evaluate scenarios to improve the productivity of R&D activities as the first step to promote innovation processes.

Chapter 4, 'Using Modeling and Simulation Techniques to Improve the Accountability and the Performance of Human Resource Management', written by Florea Nicoleta Valentina and Manea Marinela Daniela, examines whether direct costs of human capital influence on the capacity of individuals and organization to obtain motivation, work satisfaction and performance. For this purpose, it discusses the role of modeling and simulation techniques in human resource management (HRM) accountability, uses Monte Carlo simulation and compares it with three considered techniques of system dynamics, agent-based modeling, and discrete event simulation technique.

Chapter 5, 'Improve Center of Sterilized Material Performance: A Discrete Event Simulation Model', written by Daniel Bouzon Nagem Assad, Thais Spiegel, and Ana Carolina Pereira de Vasconcelos Silva, investigates Center Sterilized Material (CSM) strategy operation in a public hospital. It aims to provide a better understanding of the system behavior, the bottleneck resource and its results terms of, waiting and total time of each entity using discrete event simulation.

Chapter 6, 'Capacity Decision in Emergency Hospital Operation Rooms: Sizing Using Simulation', written by Ana Carolina Pereira de Vasconcelos Silva, Daniel Bouzon Nagem Assad, and Thais Spiegel, uses a discrete event simulation to set the amount of operation rooms needed for patient care in an emergency department. The method used can support the decision to reconfigure the allocation of emergency rooms to the installed capacity of an emergency unit based on historical demand.

Chapter 7, 'Casual Modeling to Improve Customs Services in Iran: A System Dynamics Study', written by Alireza Hassanzadeh, and Hoda Eskandari, presents a causal model to improve customs services using the methodology of system dynamics.

Chapter 8, 'A Model for the Establishment and Development of Telework in Public Organizations: Using the System Dynamics Approach', written by Alireza Hassanzadeh and Robabeh Sadat Emami Meybodi, aims to find a suitable and practical model for the establishment and successful development of telework in the public organizations using system dynamics methodology.

This book presents different case studies about applications of simulations in the areas of business process improvement, business decision making, and technology and innovation management. It contributes to impact audiences to involve in a more dynamic and effective way than traditional methods which are usually static. It provides a realistic understanding of the business environment and allows for the practice that impacts on business results. Therefore, it helps managers and decision makers make better decisions.

Simulation is considered as a tool for change management. Since simulation clarifies the necessity of change, it provides practitioners in business process management the crucial factors that lead the organizational change. Simulation enables practitioners and researchers to trace from cause to effect, and generate description of each decision and also its consequences before an action. Mental models and viewpoints of experts provide information to modelers and simulators on how processes, activities, and interactions exist and work and how to formulate strategies and make scenarios that will help managers and decision makers forecast changes caused by those decisions and their implementations. These extracted mental models and business rules have to be validated that ensure experts about the robustness of the model and results.

Scenario planning provides an opportunity to challenge business rules against external landscape within which an organization operates, so as to promote informed and enhanced decision-making. It draws a picture of the future developments of a given system in terms of interlinked elements working toward promotion of a better performance and enhancement of effectiveness. As an important supporting tool for governments, industries and communities, scenario planning helps them getting ready and planning for the future, managing risks and utilizing possible opportunities. Companies can utilize scenario planning as a method to either generate or consider transformational changes.

The increasing use of computer simulations reflects a growing desire to develop more complicated models capable of analysing the complex phenomena of technological innovation. Generally, managers, practitioners, and researchers have an imperfect understanding about the environment in which a company works, or even within a company. Therefore, modeling and simulation helps one to conceptualize a system under study which operates under a number of different circumstances, then understanding the basic structure of the system is assumed. At theoretical aspect lies concepts and the conceptual model which is very important and its influence on perception of the system is evident. However, practice in solving problems and improving system must be emphasized which is considered particularly in this book.

Agent-based modeling results not only establish useful insights into the system behavior, but it also addresses invisible interactions, especially those leading to unintended outcomes. Beside making reality more visible, using agent-based modeling makes it possible to see unexpected changes and emergent patterns which are generated by the interactions of individual entities. Emergent phenomena are difficult to understand and predict, but by using this technique, seeing unexpected events such as technological innovation is possible. Agent-based modeling provides a flexible framework to answer the question related to different fields of social systems such as an organization and its environment. The first chapter of this book uses this technique focusing on the effect of underperforming R&D alliances on innovation performance. There are many studies that have investigated innovation processes with agent-based modeling. However, the impact of continuation of underperforming alliance on performance over technology development in the pre-market stage is not addressed in any of these studies.

In order to develop agent-based systems by using agent paradigm, there are various existing application domains such as software technology, robotics, and complex systems. Multi-agent decision systems and multi-agent simulation systems are recognized as two main multi-agent systems paradigms. One of the prominent characteristics of multi-agent decision system is that agents can participate in the system to make joint and participative decisions as a group. The second chapter is about designing a multi-agent system in the case of technology strategic planning. There are some techniques and mechanisms which are used for development of marketing strategies and technology strategies, but they are different from the research of this chapter.

System dynamics is a computer-aided approach for the aim of policy analysis and design as well that is applied to solve problems of complex systems such as organizations and industries. In other words, it is helpful to know dynamic systems which have special features such interdependence, mutual interaction, information feedback, and circular causality. System dynamics is useful to study policy issues or problems to address additional questions. System dynamics is applied to study problems that arise in closed-loop systems. It is not possible to understand behavior generated by complex organizations through examining the parts. In this book, three chapters are about the applications of system dynamics. The third chapter focuses on R&D process which is mostly about intangible and have to some extent a random probability pattern. It considers R&D as a production function of material goods. In this case, inputs are the allocated resources such as budget, number of human resources or laboratory equipment. The number of innovations or patents can be recognized as output. This chapter uses system dynamic to propose scenarios of improving the productivity of R&D process by considering technical knowledge development as an indicator. There are some studies regarding using system dynamics in the analysis and scenario planning of dynamic technology innovation and R&D process which are different from the investigation of this chapter. The seventh and eighth chapters illustrate causal loop modeling of two cases respectively, in improvement of customs service, and establishment and development of telework in public organizations. In these both chapters, system dynamics is applied in four steps of: identify system's scope, extract and compile a dynamic hypothesis, build and present causal model, and confirm the model by experts.

The discrete event concept refers to a range of events which cause state changes in the system objects. It has some benefits such as presenting powerful and dynamic simulation models to replicate the system, and precise results for making better business decisions. The fifth chapter investigates CSM strategy operation in a public hospital in three steps: process description, measurement of each phase and estimate the demand based on historical data. It attempts to understand the system behavior, the bottlenecks resources and its results in terms of waiting and total time of each entity using discrete event simulation. Discrete event simulation has been widely used in modeling health-care systems in many different units like operation room, critical care units, laboratory and pharmacy, but it was found no research involving CSM and discrete event simulation. The sixth chapter applies the discrete event simulation approach to evaluate the capacity of a surgical center. In particular, in the healthcare sector, simulation is used in a large variety of studies including capacity and hospital bed planning, intensive care, operation room configurations, and time of waiting. In addition, simulation has been used for a number of purposes, since it allows to describe the system in a high degree of detail and to deal with various sources of uncertainty, without simplifying hypotheses that are otherwise necessary to obtain forecasts. Thus, this chapter aims to use discrete event simulation model to analyze the trade-off between the scenarios of dedicated rooms availability to the of emergency patients care.

Having to know the future and the potential advancement of some processes over time and their performance make organizations apply simulation, especially because using simulation for decision making is very cost-effective. It is essential in a situation when it is not possible to experience in a real system. The fourth chapter states that human capital is a crucial resource for organization and tries to find a relationship between the money spend on human capital (compensation) and the results they obtained. It uses Monte Carlo simulation-Ordinary Least Squares (OLS) method and compares it with three simulation techniques of systems dynamics, discrete event simulation, and agent-based modeling. Human resource-related accounting and finance implications are usually captured in accounting and finance texts in an unconnected manner. Throughout this chapter, the compensation and benefit topics that have major accounting implications are discussed, and referred to human capital.

Since Simulations are able to prevent the catastrophic failures in the system made by the impact of a change, it has a great importance for business analysts and policy makers. Simulations make it possible for organizations to examine new changes, procedures, information flows and the others without interrupting the smooth working of real systems. In this book, simulation models are developed to study the working of systems as they evolve over time. A fully developed and validated model can answer a variety of questions about real systems. This book was authored by leading experts to present an in-depth description of modeling and simulation theories and applications in organizations. The objective of this book is to create a comprehensive coverage of all important areas of business and enterprise from the perspective of three techniques: system dynamics, agent-based modeling, and discrete event simulation. Universities and industries are interested in this book for their related learners and experts.

## REFERENCES

Maidstone, R. (2012). *Discrete event simulation, system dynamics and agent based simulation: Discussion and comparison*. Retrieved from https://www.researchgate.net/publication/306157680_Discrete_Event_Simulation_System_Dynamics_and_Agent_Based_Simulation_Discussion_and_Comparison

# Chapter 1
# Innovation Success Over Time of Alliances With Different Strategic and Cooperation Objectives

**Khrisna Ariyanto Manuhutu**
*University of Twente, The Netherlands*

**Ariane von Raesfeld**
*University of Twente, The Netherlands*

**Peter Geurts**
*University of Twente, The Netherlands*

## ABSTRACT

*In response to uncertainty of prospective technologies and how they might fit market demand, firms tend to establish R&D alliances. In this chapter the effect over time of continuation of underperforming R&D alliances on innovation performance during the pre-market stage is investigated. This stage is characterized by non-linearity, as expected outcomes and market demands are uncertain. Literature suggests that computational modeling in particular agent-based modeling can be used to investigate such non-linear processes. Agent based modeling starts with simple behavioral rules that develop into emergent system-level behaviors, and in that way controlled system level experiments are used to identify in an inductive way causal mechanisms that drive the system development. In this chapter's simulation model, an agent decides to continue its R&D alliance based on its strategic and cooperation objectives. After evaluating if the strategic goals is met, firms can decide about the extent to which to continue the R&D alliances if the strategic goal is not met. This is called persistency.*

DOI: 10.4018/978-1-5225-3226-2.ch001

*The model is aimed to explain developmental paths and patterns of the co-evolution of alliances and technology. Despite suggestions to investigate non-linear processes in the pre-market phase by using an agent-based model, agent-based models so far do not focus on the impact of alliance continuation on innovation performance over the path of technology development. In previous research these paths mainly have been investigated in case and cross sectional studies but not in an agent based model. A base-line model is developed and the extent to which it reflects reality is analyzed in order to improve the model's performance.*

## BACKGROUND

How does continuation of an underperforming alliance affect performance over time in the pre-market stage? This stage is characterized by high uncertainty of how prospective technology can be applied and fitted to market demand (Ven, Polley, Garud & Venkataraman, 1999; Chesbrough, 2004). This circumstance triggers firms to establish R&D alliances (Tether, 2002; Hagedoorn & Duysters, 2002; Vilkamo & Keil, 2003; Faems, Looy, & Debackere, 2005; Chesbrough, 2006). A managerial problem in this R&D alliance is the decision to continue or stop underperforming alliances. Despite the importance of this decision, literature provides different explanations about the performance effect of repeated collaboration. Some empirical studies find a positive effect of repeated collaboration in alliances on performance (Doz, 1996; Reuer & Zollo (2005); Robson, Schlegelmilch, & Bojkowszky, 2012; Raesfeld, Geurts, & Jansen, 2012), and others assume a negative performance effect of repeated collaboration as it locks a firm into a current alliance and simultaneously lock the firm out of alternative alliances that might be more promising to develop new technology (Gulati, Nohria, & Zaheer, 2000). In line with the inverted U shaped effect of embeddedness (Uzzi, 1996; Hagedoorn & Frankort, 2008), performance effects of repeated collaboration might have an optimum, which might explain findings of both negative and positive effect of repeated collaboration.

So far most of the alliance literature treats alliance continuation as a dependent variable (e.g. Hoffmann, 2007; Bierly & Coombs, 2004; Doz, Olk, & Ring, 2000; Lhuillery & Pfister, 2009; Pangarkar, 2009; Reuer & Zollo, 2005; Uzzi, 1996. Among those that treat alliance continuation as an independent variable, only a few focus on the investigation of underperforming alliances, Doz (1996) and Robson, Schlegelmilch, & Bojkowszky (2012) investigate the impact of persisting in underperforming alliance but they did not investigate this over the stages of technology development.

In a nut shell, to deal with high uncertainty in a pre-market innovation, firms should form R&D alliances. And managing the continuation of the alliance can be difficult since alliance continuation can have a negative as well as a positive impact on firm performance. Moreover, research so far did not investigate the reason behind different impacts. This chapter aims to explain the different impacts of alliance continuation on innovation performance in the first stage of technology development.

## THEORY AND HYPOTHESES

The impact of alliance continuation on innovation performance will be investigated. The focus on stages of technology development originates from the findings of Gilsing & Nooteboom (2006) and Hoffman (2007) about the dynamics in alliance characteristics in these stages.

Gilsing & Nooteboom (2006) investigated network characteristics in the different stages of technology development. They presented the dynamics of the characteristics (Table 1). Among these characteristics there are some aspects that are related to alliance continuation, e.g. loose vs. formal alliances; flexible vs. formal ties; short vs. long term alliance duration. They show that in the exploration stage firms tend to establish loose and flexible alliances for a short duration. Opposite to the exploitation stage where firms tend to establish formal durable alliances.

In a similar way, Hoffmann (2007) investigated alliance portfolio parameters in different stages of technology development. He divides the exploration stages into two stages, e.g. an adapting and a shaping stage. Hoffmann found that firms tend to establish alliances with different characteristics over the stages. Some parameters can be used to proxy alliance continuation, i.e. alliance stability, linkage intensity of alliances, and number of alliances. He found that firms tend to establish many weak and unstable alliances in the adapting stage. Entering the shaping stage they more focus on a few prospective partners and nurture a stronger relationship with them. Nevertheless, due to high uncertainty, these alliances still have a high chance to end. Eventually, in the stabilizing state, the uncertainty is reduced and the prospective technology has emerged, hence they can focus on particular technology with stable partnerships since this stage is an entrance to the established market stage.

Gilsing & Nooteboom (2006) and Hoffmann's findings (2007) support each other. Their findings show similar patterns, such as that in later stages of the innovation journey, firms tend to establish stable relationships with selected partners. This tendency may reflect the relationship between these alliance characteristics and the performance in different technology development stages. This is supported by Raesfeld, Geurts, & Jansen (2012). They found that in R&D cooperation, the

*Table 1. Network characteristics in different stages of technology development*

| Exploration | | Exploitation |
|---|---|---|
| | **Governance** | |
| Radical innovation | | Incremental innovation |
| Technology oriented | | Product and process oriented |
| Experimentation with novel combination | | Experimentation in organization |
| Tacit knowledge | | Codified knowledge |
| | **Competence** | |
| Spin-offs, new entrants | | Entrance by incumbents |
| Loose alliances | | Formal alliances, acquisitions |
| Limited use of contracts | | Contracts |
| Relation-based trust | | Institution-based trust |
| | **Networks** | |
| Dense, open networks | | Non-dense, more exclusive networks |
| Informal, flexible ties | | Formalization |
| Limited size, high entry and exit | | Stabilization |
| Locally embedded | | Delocalized |
| | **Strength of Ties** | |
| High frequency of interaction | | Low frequency of interaction |
| Short duration | | Long duration |
| Higher openness | | Limited openness |
| | **Transitional Process** | |
| Divergence in knowledge and organization | | Convergence in knowledge and organization |
| Variety through break-up of existing networks and new relations to outsiders | | Selection by institutional environment |

Gilsing & Nooteboom (2006)

*Table 2. Alliance portfolio characteristics in different stages of technology development*

| Configuration Parameters of Alliance Portfolio | Adapting Strategy | Shaping Strategy | Stabilizing Strategy |
|---|---|---|---|
| Number of alliances | Many | Rather few | Few |
| Linkage intensity of alliances | Weak | Rather strong | Strong |
| Dispersion of alliances | High | Rather low | Low |
| Redundancy of alliances | Low | Rather high | High |
| Stability of alliances | Low | Average | High |

(Hoffmann, 2007)

effect of user interaction and network stability was lower for the development of applications compared to the generation of revenues. Therefore, it is reasonable to assume that in the early stage of technology development firms tend to establish weak and unstable relationships with many partners since this kind of characteristics leads to the highest innovation performance in comparison to the other strategies. In the intermediate stage of technology development firms will focus on a few prospective partners and will nurture a closer relationship with them, which will generate the highest innovation performance in comparison to those that adopt other strategies. With a similar reasoning, in the last stage of technology development, the highest innovation performance would be obtained by those that focus only on a few and stable alliances. This chain of argumentations leads to:

**Hypothesis 1:** In the early stage of technological development in which market uncertainty is high and firms explore new innovation opportunities, the degree of repeated collaboration in the same alliances has a negative impact on performance.

**Hypothesis 2:** In the later stage of technological development in which market uncertainty is average, and firms start to exploit the innovation opportunities, the degree of repeated collaboration in the same alliances has an inverted U shaped impact on performance.

**Hypothesis 3:** In the stage of exploitation when market uncertainty is low, the degree of repeated collaboration in the same alliances has a positive impact on performance.

In this paper the effect of the extent of continuation of underperforming alliances on emerging performance in dyadic inter-organizational cooperation will be investigated at the field of R&D projects in which a combination of diverse industry partners collaborate. Performance is measured by the fit to the market demand. This is executed by varying the persistency or degree of repeated collaboration based on previous cooperation experience under different strategic goals. For this purpose an agent-based model is developed. The properties and behavior of the agents and their environment in this model stem from the descriptive results of case study research on managing strategic alliances (Gilsing & Nooteboom, 2006; Hoffman, 2007).

## WHY AGENT-BASED MODELING?

To investigate the impact of continuation of underperforming alliances on the innovation performance in the pre-market stage of technology development, agent-based modeling will be used. This approach is used for the following reasons: to be able to study the non-linear impact between action and outcome in an innovation process (Van de Ven & Poole, 2005), to be able to explore what might be beyond realm (Burton & Obel, 2011), and to be able to isolate the observed phenomenon.

The first reason to use agent-based modeling is to study non-linear processes. Van de Ven and Poole (2005) suggest to study non-linear processes in innovation development using agent-based modeling. The pre-market stage is characterized by a non-linear relationship between action and outcome of the action (Cheng & Van de Ven, 1996; Van de Ven, Polley, Garud, & Venkataraman, 1999). This non-linearity can be reflected in situations in which: 1. small changes in action can lead to drastic changes in outcomes and; 2. the same change of action may lead to different outcomes at different times. The first situation is called epistatic by Kauffman (1989) and Levinthal (1997). Epistatic situations emerge as a consequences of interaction between various actions in the innovation process. Hence, a change in one action can influence the fitness of the other actions strongly. The second situation, in which the same change of action can lead to different outcome at different times can be caused by asynchronous development between complementary technologies. This asynchronous development may lead to diachrony, a situation where an outcome of technology development may vary over stretches of time (Garud, Tuertscher & Van de Ven, 2013). According to Van de Ven & Poole (2005), conventional variance and process methods have difficulties to deal with non-linearity. Variance methods cannot capture the interaction embedded in the innovation process, while, the process methods cannot explain the consequences of the interaction in the innovation process.

The second reason to use agent-based modeling is that this tool gives the opportunity to explore what might be beyond realm. According to Burton and Obel (2011) computational modeling, e.g. agent-based modeling is a laboratory where researchers can extend their experimentation and reach beyond describing and explaining what is, towards exploring what might be. Due to high uncertainty of outcomes in the pre-market stage, particularly in the early stage, there is a lack of observations in the real world about continuation of underperforming alliances. Hence, the use of agent-based modelling in this investigation possibly provides understanding beyond what exists in reality.

The final reason to use agent-based modeling is that it permits isolation of phenomena under scrutiny. Unlike natural phenomena, social phenomena are characterized by the researcher's inability to isolate and repeat. Thus, the definition of science in which one should be able to isolate and repeat cannot be applied here in the social and behavioural sciences. Although the real social phenomenon cannot be isolated and repeated, the use of agent-based models permits the creation of artificial societies. In such societies, researchers will be able to isolate and repeat artificial social phenomena. The isolation and repetition is helpful to analyze the relationship between action and outcome in the addressed issue.

The existing agent based models are aimed to investigate a variety innovation phenomena, such as, the impact of universities existence on network performance (Ahrweiler, Pyka & Gilbert, 2011). Further, Schrempf and Ahrweiler (2012) offered a modeling concept of technology development in innovation network for general-purpose technologies. Ma & Nakamori (2005) proposed an agent-based model that simulates the process of technological innovation. Muller, Buchmann & Kudic (2014) investigated the impact of partner choice strategy on network structure. Korber, Paier & Fischer (2009) investigated the effect of different type of public Research, Technology, and Innovation (RTI) funding on innovative performance. Aggarwal, Siggelkow & Singh (2011) and Llerena & Ozman (2013) modeled the effect of mode of alliance governance on firm's or alliance's performance. Ohori & Takahashi (2012) examined scenario analysis in the competition between new technologies. Carayannis, Provance & Givens (2011) investigated the role of absorptive capacity (knowledge acquisition and transformation) on the sustainability of new venture formation in a region So far these studies, do not investigate the impact of alliance continuation on innovation performance over the stages of technology development.

## THE AGENT-BASED SIMULATION MODEL USED IN THIS STUDY

The basic construction of the model of this investigation is inspired by the SKIN model, although the mechanisms investigated in the proposed model are different to that of the SKIN model. The SKIN model is developed by a group of researchers (Korber, Paier, & Fischer, 2009; Ahrweiler, Pyka, & Gilbert, 2011; Schrempf & Ahrweiler, 2012; Muller, Buchmann, & Kudic, 2014) to model innovation development taking place in industry university cooperation.

This investigation continues on the initial idea of the SKIN model since the model represents two important aspects of technology development in the exploration stage, e.g. recombination of existing technology and non-linear exploration behavior in the landscape. Related to the first aspect, some studies (Fleming & Sorenson, 2001; Enkel, Gassmann, & Chesbrough, 2009; Geisendorf, 2011) show that, in the pre-market stage, most of the new technologies (knowledge) are generated through recombination of existing technologies (knowledge). The way the existing technologies are recombined with each other resembles that of genetic cross-over in genetic algorithms (Geisendorf, 2011). Accordingly, she suggests a chain of genes as a good representation of technology development. The second aspect, non-linear exploration behavior, has also been incorporated into the SKIN model. The non-linear behavior is caused by a journey in a rugged landscape searching for prospective technology as well as a prospective market that fit the technology (Van de Ven, Polley, Garud, & Venkataraman, 1999). The origin of this rugged landscape concept is derived from Kauffman (1989) and Levinthal (1997). So in line with the SKIN model we represent the technological competences (knowledge and skill) of firms and the development of competences as changes of genes that are recombined in interaction. The experiment in this chapter compares the innovation performance of alliances at different levels of repeated collaboration after failure over time, under varying strategic goals. The rugged landscape is approximated by randomly assigned knowledge of the alliances at the start of the simulation.

## Description of the Agent-Based Model

With the agent-based model presented in this chapter, markets are simulated in order to study if and in how far strategic goals and strategies concerning the decision to continue or to stop with a dyadic alliance aimed at developing an innovative product influence the performance of the alliance.

A market consists of firms collaborating in dyadic alliances aimed at developing technology. The product of the innovative effort is called innovation offering. The better the fit of the offering with the demanded innovation the better the innovation performance of the alliance is.

Assumed is that a large variety of knowledge exists. Some knowledge may not yet be discovered but in this abstract market it is assumed that some still have to be discovered but are potentially already there (see the gene combination reasoning). In aiming at innovation it is not necessary to discern between already existing knowledge and still to be developed knowledge and it is assumed that all knowledge is already potentially there. A knowledge element is called a competence. Assumed is here that in a market a specific combination of competences is demanded. The innovation performance of an alliance is better when the overlap of the offering with the demand is larger.

It is assumed further that each firm has a limited number of competences, which is a sample of all possible competences in the market.

A part of the competences of a firm is brought into the alliance.

The combined competences of the part of two firms brought into the alliance form the alliance proposal and form as such the alliance competence.

A selection from the alliance competences forms the innovative offering.

The innovative offering is compared with the demand, the overlap between the two is the market fit. In this way firms learn via their alliance what is an innovative and what a failed competence.

The partners in the alliance expect as a strategic goal a minimum performance of the alliance, expressed in the market fit, in order to continue the alliance. After a certain time the performance of the alliance is evaluated, again expressed in market fit. In case the alliance performed well, i.e. the actual market fit is equal to or exceeds the strategic goal, the alliance is continued.

In case the performance expectations of the alliance are not met, the partners may decide not to continue the alliance and start a new alliance with another firm. They may also decide that they continue with the alliance in spite of its underperformance. The number of times the alliance partners sequentially decide to continue is called persistency level or alliance continuation strategy (Figure 1).

In combining competences of both firms into the alliance, the firms learn from each other. Through the alliance the firms learn about the earlier unknown competences of their partner.

Distinction between phases of exploration and exploitation is made by number of years over time (early stage of exploration 1-5 years; later stage of exploration, 5-10 years; and exploitation after 10 years and onwards).

The uncertainty of the innovation market (how rugged the innovation landscape is) is expressed as the mean market fit of the offerings of all alliances at a certain point in time.

To get observations regarding the performance of different alliance continuation strategies, parallel isolated artificial societies or innovation markets are introduced (Figure 2), or in other words various experimental conditions are applied. In the first series of experiments, these artificial societies are identical in all aspects. A firm (an agent) in a society has equals in all other societies. These equals possess exact similar competences in their knowledge base. They are also connected to the same partners.

In a second and more complex series of the same experiments we simulate 10 societies with different starting configurations of firm's competences and assignment of alliance partners. Moreover, strategic goal and persistency level are varied. In this way we are able to assess to what extent the starting configuration, or typical form of the rugged landscape in a particular society has impact on the innovation performance.

## Experimental Setting

We define the initialization of the model (Figure 3) as follows:

- In the first series of experiments there is one persistency level (0). In the second series of experiments there are 10 persistency levels (0 to 9).

Ten persistency levels are chosen to create an observation range. If the number is too small, e.g. one or two level of persistency, the use of agent-based may be

*Figure 1. The concept for comparing the performance of firms that adopt different persistency level*

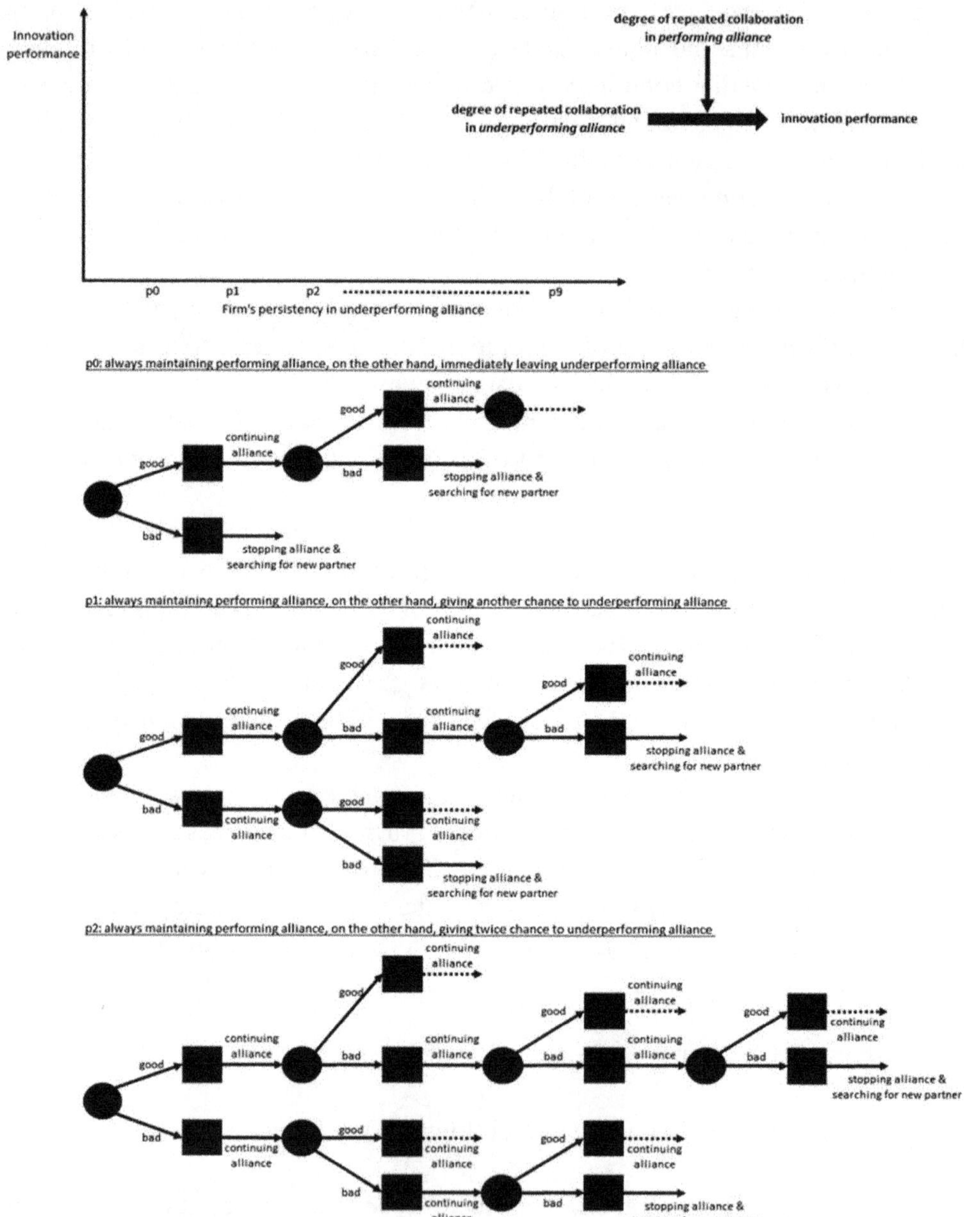

*Figure 2. Global layout of the agent-based model*

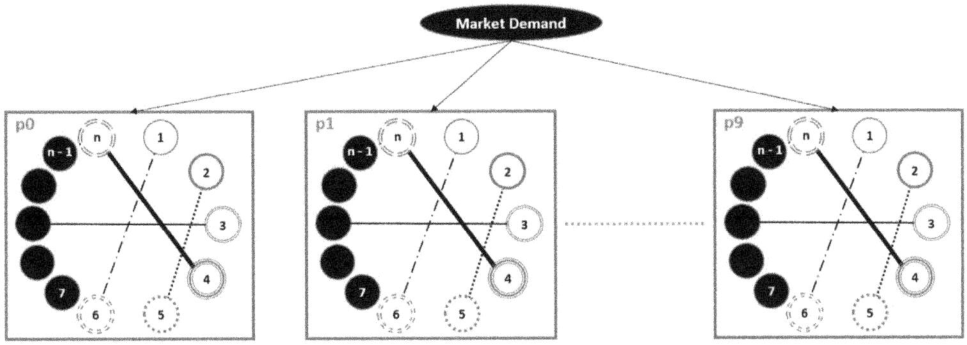


questioned since such data might be obtained from the real world. On the other hand, if persistency level range is set too high e.g. 20, it might be too far from reality.

- 100 firms, and therefore 50 alliances. Each firm always has precisely one alliance at a time. With 50 alliances we create so much variation in the genetic combinations that we may apply the law of the greater numbers.
- In the first series of experiments the strategic goal is set on 100% market fit. In the second series of experiments the strategic goals are varied in order to simulate the phases in the innovation journey. The later in the journey the higher the performance expectations. The model applies 0, 20, 40-60 80 and 100% market fit as strategic goals.
- At each strategic goal level all 10 persistency levels are applied. A persistency level is made operational as the number of times that firms in sequential periods of time decide to continue the collaboration in the alliance despite not reaching their strategic goal.
- 500 technology competences are in the pool.

The setting of the total possible competence at 500 competences is to present a simple innovation model. Enlarging this number will only lead to a slower innovation trajectory, and vice versa, the patterns won't change. 20 Potent technology competences are in each firm's knowledge base, (randomly selected from the 500 technology competences in the knowledge pool). Again a larger knowledge base will lead to quicker learning making higher persistency levels more attractive for innovation, and vice versa.

- These 20 competences are initially all potentially innovative or potentially failing.
- A competence is failing if there is no match with the demand
- A competence is innovative if there is a match with the demand
- Market demand, a combination 12 technology competences (randomly selected from the 500 competences in the knowledge pool).

Collaboration with a new partner and repeated collaboration with the actual alliance partner and innovation decision are defined as follow:

- Collaboration with a new partner
    - Firms randomly choose an alliance partner.
    - The firm's competences are a list of innovative, failing and potent competences. The firms exclude the failing competences as eligible for this actual and later alliance proposals.
    - Firms add their innovative (fit) competences to the firm's alliance proposal.
    - If a proposal has reached the maximum length, in this case 12, it is sent to alliance pool. Otherwise, the shortage should be filled from firms' potent competences. The proposal filled to the maximum is then sent to the pool.
    - The alliance proposal is now filled with 24 innovative and potent competences. 12 from these 24 are selected randomly from the alliance proposal and assigned to the alliance offering (new technology). Competency ties in the offering are taken care of by random replacement from the firm's alliance proposal into the alliance offering.
    - The alliance offering and the market demand is matched.
- Repeating the collaboration in the alliance
    - The firm's competences are in a list of innovative, failing and potent competences. The firms exclude the failing competences as eligible for an alliance proposal.
    - Firms assign their successful (fit) competences to the firm's alliance proposal.
    - If the new firm's alliance proposal has reached the maximum length (12) this proposal is sent to the alliance proposal pool. Otherwise, the shortage should be filled from firms' potent competences. The filled proposal is then sent to the pool.

    - 12 From the 24 competences are selected randomly from the alliance pool and assigned the alliance offering (new technology). Competency ties in the offering are taken care of by random replacement from the firm's alliance proposal into the alliance offering.
    - Match the offering and the market demand.
- If the alliance's offering surpasses the strategic goal, they continue collaboration. Otherwise, they reduce persistency level by one. As long as persistency level is higher than zero, the alliance will continue. If not, they will stop collaboration and choose for the next period a new alliance partner (see collaboration with a new partner).
- All innovative competences that has not been listed yet in in the list firm competences will be added in exchange with failed competences. In this way the firms in an alliance learn from each other.
- If the number of new innovative competences is larger than the number of failed competences, a randomly chosen potent competence is removed from the firms list of competences until a list of 20 firm competences remain.
- For the next year of collaboration all innovative competences are assigned to the alliance proposal, a random selection of potent competences is added. In this way the learning is cumulative through time.
- The experiment is repeated 20 times representing 20 years.

The flowchart of collaboration is displayed in Figure 4 and 5.

## Decision Diagram (Flow Chart)

See Figures 3-5.

*Figure 3. The initialization of model*

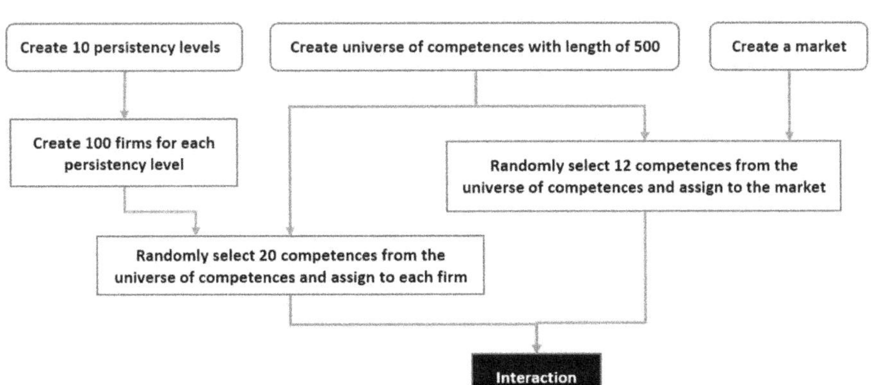

*Figure 4. The collaboration and innovation decision rule of the model*

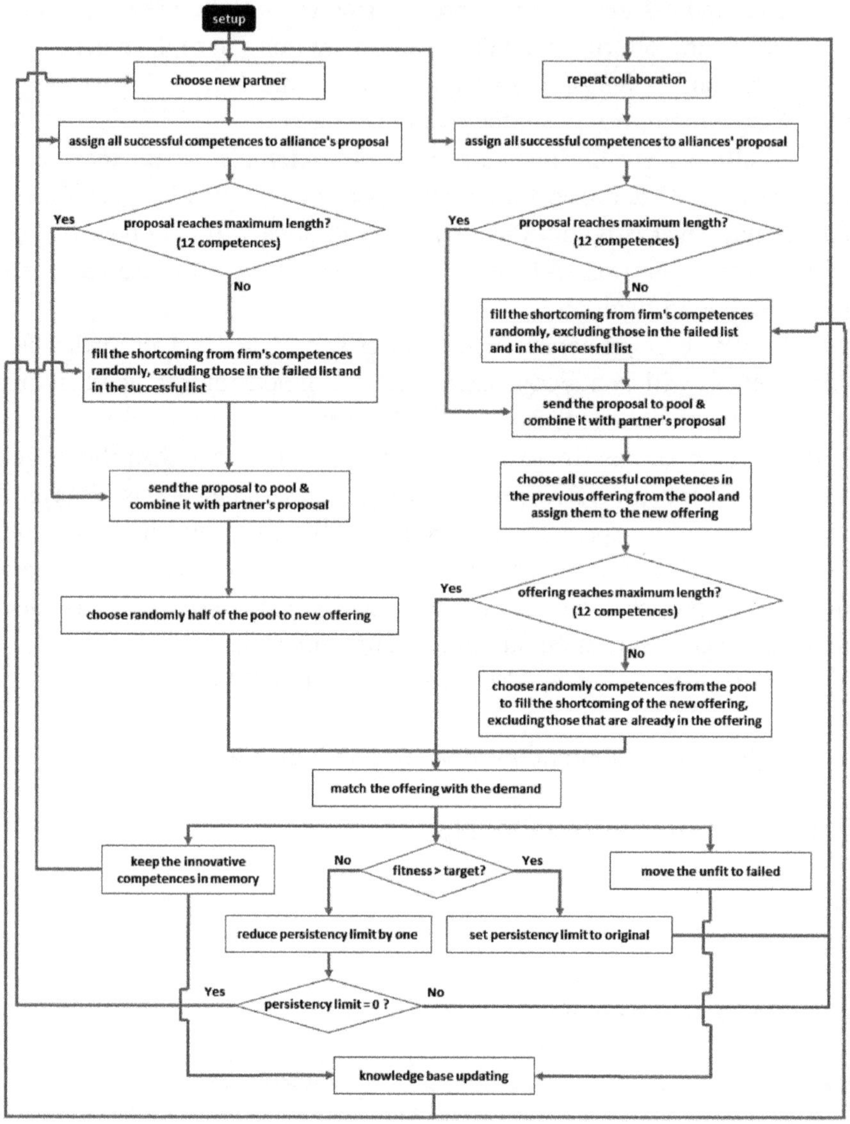


## RESULTS

The results of the first series of experiments with 20 experimental conditions (one society, persistency level 0, one strategic goal over twenty years) are presented in Figure 6 and 7.

*Figure 5. The knowledge updating part of decision flow chart*

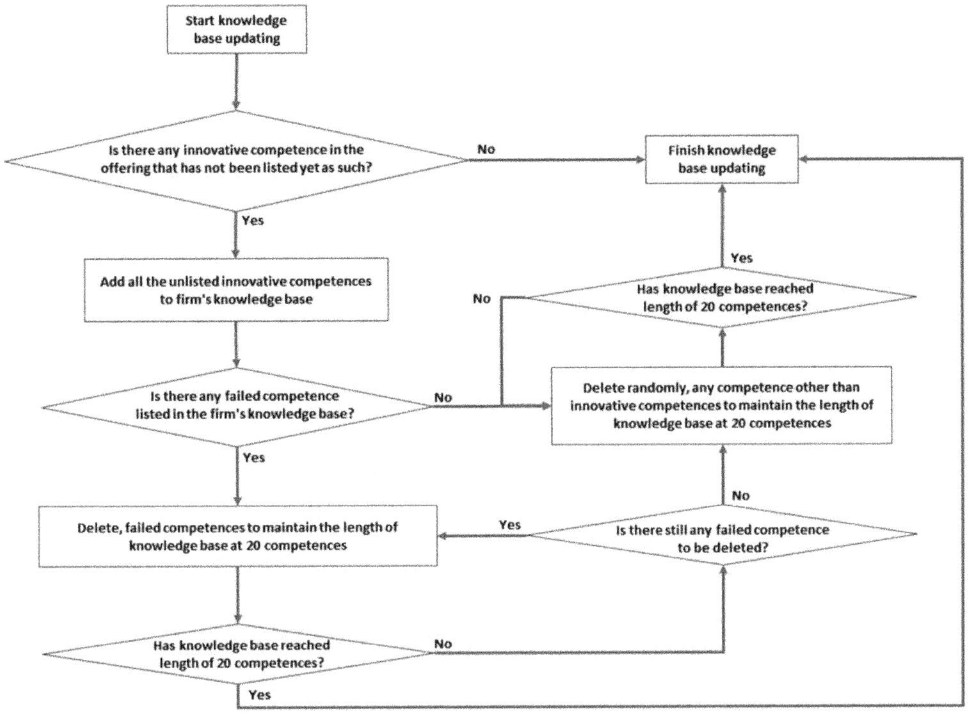


*Figure 6. Market uncertainty (100 – mean market fit) at persistency level 0 over time for strategic goal 100% market fit in one society*

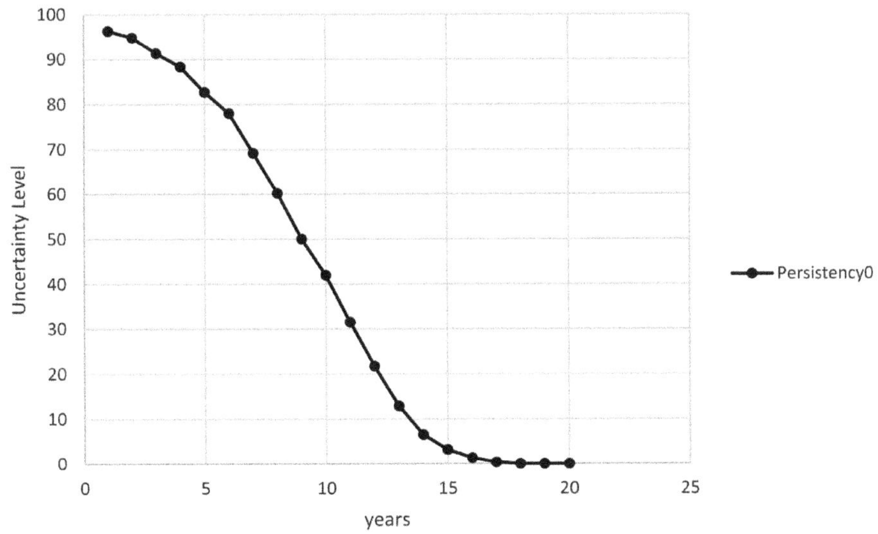

*Figure 7. Average number of successful firms per year with 100% market fit at persistency level 0 for strategic goal 100% market fit*

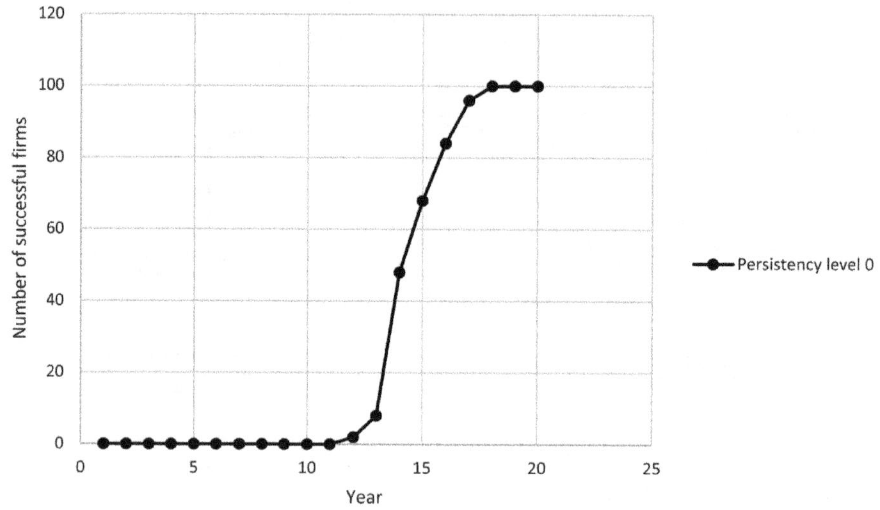


Under the experimental conditions the curve for persistency level 0 in Figure 6, shows the system level market uncertainty (100 – mean market fit) over time this reflects the three stages of technological development from year 0-5 the early stage of exploration with high market uncertainty, from year 5-10 the later stage of exploration with medium market uncertainty, and from year 10- 20 the exploitation stage with low market uncertainty.

Figure 7 shows the firms that reach a 100% market fit over time at persistency levels 0 for strategic goal 100% market fit. Figure 7 shows that in the first series of experiments for a strategic goal of 100% market fit and a persistency level of 0, the first successful alliance with 100% market fit gets up after 12 years, after 13 years there are four successful alliances followed by a steep growth and at 18 years all alliances are successful.

Figure 8 – 11 represent the result of a second series of experiments consisting of 10 different societies with the following conditions, ten societies, six strategic goals (e.g. respectively target market –fit of 0, 20, 40, 50, 60, 80, and 100%), and ten persistency levels over 20 year. Figure 8, shows the decrease of market uncertainty over 20 years for strategic goal 100% market-fit. Compared to the first series of experiments this decrease in market uncertainty is much slower. After 20 years there still is a market uncertainty of 65%, indicating a strong dependence on the initial configuration of the society and thus unreliability of the outcomes.

Figure 9, describes the persistency time combination in which successful alliances appear for particular strategic goals, it shows that only strategic goal 60, 80, and

*Figure 8. Average market uncertainty of 10 persistency levels over time for strategic goal 100% market fit over a sample of ten societies*

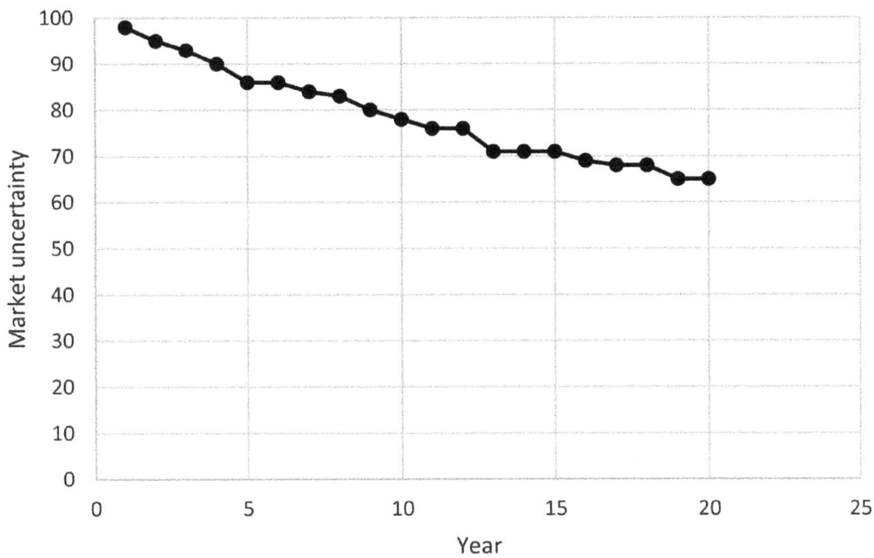


*Figure 9. Average number of persistency time combinations with successful firms of ten persistency levels over twenty year, and for six strategic goals over a sample of ten societies*

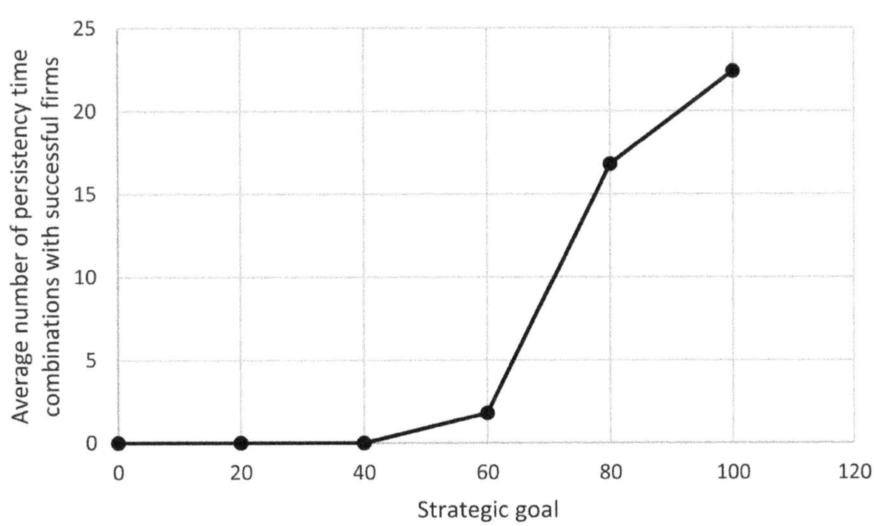

100% independently from the persistency level can lead to successful alliances. Thus, low strategic goals do not lead to success.

Figure 10 shows the effect of persistency level on success of alliances independent of time and sample and indicates that a low persistency level with a high strategic goal is the most successful.

Figure 11 shows that in the first five year the strategic goal is not important and that strategic goals lower than 60% do not lead to success. The graph for strategic goal 100% shows that after 6 years there is a 1% chance to find a successful alliance, after 11 year there is a chance of 8% to find a successful firm and after 20 years this chance increased to 30%.

*Figure 10. Average number of year with successful firms for ten persistency levels and strategic goals 60%, 80%, and 100% market fit over a sample of ten societies*

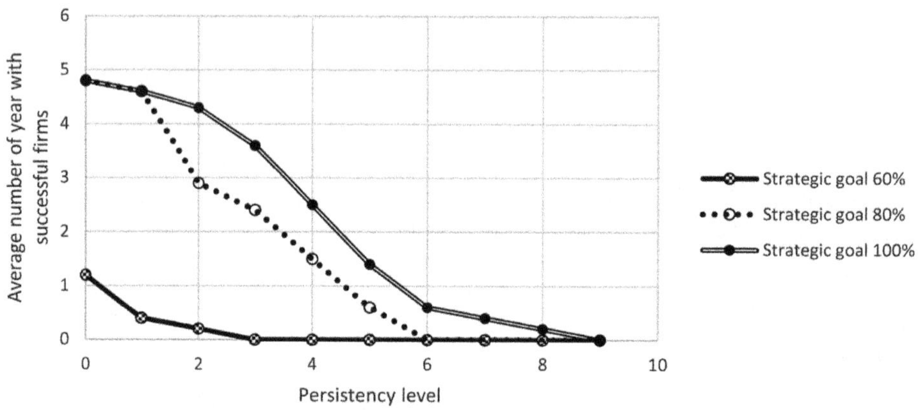


*Figure 11. Average number of persistency levels with successful firms for strategic goals 60%, 80% and 100% market fit over twenty year of a sample of 10 societies*

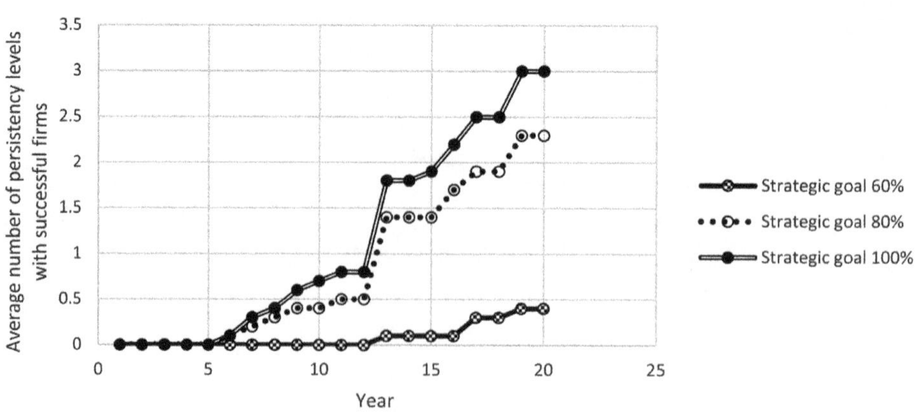

## CONCLUSION

The first series of experiments are used to show how the mechanisms in the model work or how strategic goal and persistency level impact market uncertainty. At the system level development of market uncertainty over time indicates different stages of technological development, from year 1-5 early stage, from year 5-10 middle stage, from year 10-20 late stage. In this way it is assumed that each phase lasts about 5 year. This is made operational by dividing the development years in 3 parts of 5 year each. This results in accepting Hypothesis 1: In the early stage of technological development in which market uncertainty is high and firms explore new innovation opportunities, the degree of repeated collaboration in the same alliances has a negative impact on performance. And rejecting Hypothesis 2 and 3.

In the second series with 10 societies there is a limited decrease of market uncertainty at the system level, indicating strong dependence on the initial configuration of the firms. This also implies that the experiments only show the early stage of technological development. So over the whole simulated period of 20 year the system is highly uncertain. According to this fact the system never leaves the early phase of innovation development. It also explains why Hypothesis 2 and 3 are to be rejected. According to the general system's uncertainty level Hypothesis 2 and 3 are not applicable in our simulations (a test is therefore not valid). Another objection against this operationalization of the innovation phases is that there are actual innovating firms after 6 years (see Figure 11). This means that the general systems situation is probably not a valid indication of the actual phases in which firms and alliances are in their innovation journey. With the experiments conducted in this research the first hypotheses can be confirmed. In order to test the other hypotheses a solution have to be found to simulate a decrease of market uncertainty 100 to 0 percent. A possible direction for to simulate market uncertainty is to assume that firms and alliances have little knowledge about the general systems state, but that they have knowledge about their own progress since the start of their innovation journey. So a next step is to find a solution to how alliances (firms) judge their progress in the innovation journey

## REFERENCES

Aggarwal, V. A., Siggelkow, N., & Singh, H. (2011). Governing collaborative activity: Interdependence and the impact of coordination and exploration. *Strategic Management Journal*, *32*(7), 705–730. doi:10.1002/smj.900

Ahrweiler, P., Pyka, A., & Gilbert, N. (2011). A new model for university-industry links in knowledge-based economies. *Journal of Product Innovation Management, 28*(2), 218–235. doi:10.1111/j.1540-5885.2010.00793.x

Bierly, P. E. III, & Coombs, J. E. (2004). Equity alliances, stages of product development, and alliance instability. *Journal of Engineering and Technology Management, 21*(3), 191–214. doi:10.1016/j.jengtecman.2004.05.001

Burton, R. M., & Obel, B. (2011). Computational modeling for what-is, what-might-be, and what-should-be studies-and triangulation. *Organization Science, 22*(5), 1195–1202. doi:10.1287/orsc.1100.0635

Carayannis, E. G., Provance, M., & Givens, N. (2011). Knowledge arbitrage, serendipity, and acquisition formality: Their effects on sustainable entrepreneurial activity in regions. *IEEE Transactions on Engineering Management, 58*(3), 564–577. doi:10.1109/TEM.2011.2109725

Cheng, Y.-T., & van de Ven, A. (1996). Learning the innovation journey: Order out of chaos? *Organization Science, 7*(6), 593–614. doi:10.1287/orsc.7.6.593

Chesbrough, H. (2004). Managing open innovation. *Research Technology Management, 47*(1), 23–26.

Chesbrough, H. (2006). Open innovation: a new paradigm for understanding industrial innovation. In H. Chesbrough, W. Vanhaverbeke, & J. West (Ed.), Open Innovation: Researching a New Paradigm. Oxford University Press.

Doz, Y. L. (1996). The evolution of cooperation in strategic alliances: Initial conditions or learning processes? *Strategic Management Journal, 17*(S1), 55–83. doi:10.1002/smj.4250171006

Doz, Y. L., Olk, P. M., & Ring, P. S. (2000). Formation Processes of R&D Consortia: Which Path to Take? Where Does It Lead? *Strategic Management Journal, 21*(3), 239–266. doi:10.1002/(SICI)1097-0266(200003)21:3<239::AID-SMJ97>3.0.CO;2-K

Enkel, E., Gassmann, O., & Chesbrough, H. (2009). Open R&D and open innovation: Exploring the phenomenon. *R & D Management, 39*(4), 311–316. doi:10.1111/j.1467-9310.2009.00570.x

Faems, D., van Looy, B., & Debackere, K. (2005). Inter organizational collaboration and innovation: Toward a portfolio approach. *Journal of Product Innovation Management, 22*(3), 238–250. doi:10.1111/j.0737-6782.2005.00120.x

Fleming, L., & Sorenson, O. (2001). Technology as a complex adaptive system: Evidence from patent data. *Research Policy*, *30*(7), 1019–1039. doi:10.1016/S0048-7333(00)00135-9

Garud, R., Tuertscher, P., & van de Ven, A. (2013). Perspectives on innovation processes. *The Academy of Management Annals*, *7*(1), 775–819. doi:10.1080/19416520.2013.791066

Geisendorf, S. (2011). Internal selection and market selection in economic genetic algorithms. *Journal of Evolutionary Economics*, *21*(5), 817–841. doi:10.1007/s00191-010-0193-1

Gilsing, V., & Nooteboom, B. (2006). Exploration and exploitation in innovation systems: The case of pharmaceutical biotechnology. *Research Policy*, *35*(1), 1–23. doi:10.1016/j.respol.2005.06.007

Gulati, R., Nohria, N., & Zaheer, A. (2000). Strategic networks. *Strategic Management Journal*, *21*(3), 203–215. doi:10.1002/(SICI)1097-0266(200003)21:3<203::AID-SMJ102>3.0.CO;2-K

Hagedoorn, J., & Duysters, G. (2002). External sources of innovative capabilities: The preference for stratregic alliances or mergers and acquisitions. *Journal of Management Studies*, *39*(2), 167–188. doi:10.1111/1467-6486.00287

Hagedoorn, J., & Frankort, H. T. W. (2008). The gloomy side of embeddedness: The effects of overembeddeddness on inter-firm partnership formation. *Network Strategy*, *25*, 503–530. doi:10.1016/S0742-3322(08)25014-X

Hoffmann, W. H. (2007). Strategies for managing portfolio of alliances. *Strategic Management Journal*, *28*(8), 827–856. doi:10.1002/smj.607

Kauffman, S. (1989). Adaptation on rugged fitness landscapes. In D. L. Stein (Ed.), *Lectures in the Sciences of Complexity*. Reading, MA: Addison-Wesley.

Korber, M., Paier, M., & Fischer, M. M. (2009). An agent-based view of the biotech innovation system. *Third Central European Conference in Regional Science*.

Levinthal, D. A. (1997). Adaptation on rugged landscapes. *Management Science*, *43*(7), 934–950. doi:10.1287/mnsc.43.7.934

Lhuillery, S., & Pfister, E. (2009). R&D cooperation and failures in innovation projects: Empirical evidence from French CIS data. *Research Policy*, *38*(1), 45–57. doi:10.1016/j.respol.2008.09.002

Llerena, P., & Ozman, M. (2013). Networks, irreversibility and knowledge creation. *Journal of Evolutionary Economics*, *23*(2), 431–453. doi:10.1007/s00191-011-0231-7

Ma, T., & Nakamori, Y. (2005). Agent-based modeling on technological innovation as an evolutionary process. *European Journal of Operational Research*, *166*(3), 741–755. doi:10.1016/j.ejor.2004.01.055

Muller, M., Buchmann, T., & Kudic, M. (2014). Micro strategies and macro patterns in the evolution of innovation networks: an agent-based simulation approach. In N. Gilbert, P. Ahrweiler, & A. Pyka (Eds.), *Simulating Knowledge Dynamics in Innovation Networks*. Springer-Verlag.

Nelson, R. R., & Winter, S. G. (1982). The Schumpeterian tradeoff revisited. *The American Economic Review*, *72*(1), 114–132. doi:10.1007/978-3-662-43508-3_4

Ohori, K., & Takahashi, S. (2012). Market design for standardization problems with agent-based social simulation. *Journal of Evolutionary Economics*, *22*(1), 49–77. doi:10.1007/s00191-010-0196-y

Pangarkar, N. (2009). Do firms learn from alliance terminations? An empirical examination. *Journal of Management Studies*, *46*(6), 982–1004. doi:10.1111/j.1467-6486.2009.00844.x

Reuer, J. J., & Zollo, M. (2005). Termination outcomes of research alliances. *Research Policy*, *34*(1), 101–115. doi:10.1016/j.respol.2004.11.003

Robson, M. J., Schlegelmilch, B. B., & Bojkowszky, B. (2012). Resource deployment stability and performance in international research-and-development alliances: A self-determination theory explanation. *Journal of International Marketing*, *20*(1), 1–18. doi:10.1509/jim.11.0072

Schrempf, B., & Ahrweiler, P. (2012). Modelling innovation networks of general purpose technologies – the case of nanotechnology. *Proceedings of the 2012 Winter Simulation Conference*. doi:10.1109/WSC.2012.6465327

Siggelkow, N., & Rivkin, J. W. (2006). When exploration backfires: Unintended consequences of multilevel organizational search. *Academy of Management Journal*, *49*(4), 779–795. doi:10.5465/AMJ.2006.22083053

Tether, B. S. (2002). Who co-operates for innovation, and why: An empirical analysis. *Research Policy*, *31*(6), 947–967. doi:10.1016/S0048-7333(01)00172-X

Uzzi, B. (1996). The sources and consequences of embeddedness for the economic performance of organizations: The network effect. *American Sociological Review*, *61*(4), 674–698. doi:10.2307/2096399

Van de Ven, A., Polley, D., Garud, R., & Venkataraman, S. (1999). *The Innovation Journey*. Oxford University Press.

Van de Ven, A., & Poole, M. S. (2005). Alternative approaches for studying organizational change. *Organization Studies*, *26*(9), 1377–1404. doi:10.1177/0170840605056907

Vilkamo, T., & Keil, T. (2003). Strategic technology partnering in high velocity environments-lessons from a case study. *Technovation*, *23*(3), 193–204. doi:10.1016/S0166-4972(01)00108-0

von Raesfeld, A. M., Geurts, P., & Jansen, M. (2012). When is a network a nexus for innovation? A study of public nanotechnology R&D projects in the Netherlands. *Industrial Marketing Management*, *41*(5), 752–758. doi:10.1016/j.indmarman.2012.06.009

# Chapter 2
# BDI Approach to Build a Single Agent of a Distributed Multi-Agent System

**Maryam Ebrahimi**
*Azad University, Iran*

## ABSTRACT

*The main purpose of this study is to describe and analyze an agent from a distributed multi-agent based system (ABS) according to the BDI architecture. This agent is capable of autonomous action to propose general technology strategies (TSs) in renewable energy SMEs based on a set of rules and interacts with a core agent in multi ABS. The recognition of internal strengths and weaknesses as well as external opportunities and threats takes place on the basis of technological SWOT-analysis. Proposed TSs are categorized into four types: aggressive strategy, turnaround oriented strategy, diversification strategy, and defensive strategy. Agent architecture in terms of three abstraction layers called psychological, theoretical, and implementation is explained. And after system validation by experts, some program codes and output results of this agent are presented. This system provides information to facilitate the TS planning process to be carried out effectively.*

## INTRODUCTION

Technological competencies, amount of investment, technology sourcing and competition need to be managed based on a strategic point of view. The first and most important step of technology management is to develop a long-term strategic plan which determines firm's investment priorities and preferences in the field of

DOI: 10.4018/978-1-5225-3226-2.ch002

technology development, called "Technology Strategy" (TS) (Arasti and Packniat, 2011). TS is different from overall business strategy due to the essence of the higher uncertainty, more use of intellectual property, and facing different strategic issues (Shane, 2009).

Besides, small and medium-sized enterprises are important actors in any economy. Furthermore, SMEs are known to be an important determinant of economic growth and employment and their potential in terms of innovations is evident too (Kubíčková et al., 2014). Many smaller firms face resource constraints, and existing resources must consequently be used with care, as erroneous decisions will have more serious complications than they would have in large businesses (Edvardsson and Durst, 2013). TS in SMEs is different from large ones. There are some differences such as: the way of protecting intellectual property, R&D intensity, and approach to generating competitive advantage (Shane, 2009).

Since a range of essential activities - including agriculture, computing, manufacturing, construction, and health and social services - depends on access to energy, it is essential to consider technology to meet and support energy development as the starting point and having long-term technology plan. Renewable energy can make a substantial contribution in each of the above mentioned areas. It is in this context that the role of renewable energy needs to be seen. It is no longer alternate energy, but will increasingly become a key part of the solution to the nations' energy needs.

The increase in knowledge, technological progress, changes, and high level of uncertainties in environment pose real challenges to TS planning. Besides, complexity of strategic decisions means as inter-relationship among choices causes troublesome challenges in strategy planning. On this basis, academicians and researchers involved in strategic management have devoted increasing attention to different methods and techniques to provide support for managers and decision makers. The main purpose of this study is developing a BDI agent – based system on the basis of SWOT analysis to suggest general TSs in renewable energy SMEs. SWOT analysis as a decision matrix in TSs planning helps identify the strategies that will best align with company's resources and capabilities to the requirements of the environment in which the firm operates. For finding SWOT factors, interview and survey methods were used by asking from forty experts. System architecture, some program codes and output results are presented. WIN-PROLOG 5.0 is applied as a programming tool and the Prometheus is used for designing the agent based model. The system is validated by experts and it proposes TSs including aggressive strategy, turnaround oriented strategy, diversification strategy, and defensive strategy to support managers in the TS planning process.

## TECHNOLOGY STRATEGY METHODS

There are a wide range of methods pertinent to development of strategies for technological progress. Ghazinoory et al. (2009) referred to technology planning approach which emphasizes on the use of strategic decision matrix and usually bi-dimensional matrix of attractiveness–ability. It involves determining prime technologies, evaluating the attractiveness and firm ability dealing with identified technologies, and locating the technologies in the matrix according to the results of evaluations. On the basis of this model, four types of TSs can be proposed in each defined technology. This method focuses on both approaches of resource – based view (RBV) and positioning (industry and competition) as it concerns the external environment and the firm's competencies. Cardullo (1996) represented technological SWOT analysis as an approach to consider the inhibitors and enhancers of technology development that a firm encounters in both its internal and external environments. Thus, it regards four factors including strength, weakness, opportunity, and threat to propose four types of general TSs. This method focuses on RBV and positioning approaches.

Matrix presented by AD Little based on the strategic impact of technology and the anticipated cost of technology development respecting each technology, recommends three technology sourcing decisions (Floyd 1998). This method is according to RBV and positioning approaches. Additionally, matrix of competitive position of the technologies - competitive impact of A. D. Little's model emphasizing on positioning proposes TSs in case of each technology.

Booz-Allen and Hamilton methodology using technology matrix based on technology importance and position of technology, and business matrix according to business attractiveness and position of business, regarding positioning approach, recommends investment prioritization for each technology (Pappas 1984). Chiesa (1996; 2001) mentioned technology – application matrix to propose five TSs concerning each technology with regard to RBV approach. As it is clear from decision matrixes, all methods on the basis of decision matrices instead of technological SWOT analysis concentrate on a specified technology. Therefore, in this study, to propose general TSs in firm's level, technological SWOT analysis is adopted. SWOT is the foundation for evaluating the internal potential and limitations and the probable opportunities and threats from the external environment. It views all positive and negative factors inside and outside the firm that affect the success. As a first step in the development of a TS planning system in firms, managers should commence with the identification and evaluation of these strategic factors which assist or hinder the company in reaching its full potential (Houben et al., 1999). Existing methods related to development of strategies for TS are presented in Table 1.

*Table 1. Comparison between TS techniques*

| Method | Input | Process | Output | Reference |
|---|---|---|---|---|
| Critical technologies | Feasibility, attractiveness, and list of areas of technologies | Feasibility – attractiveness matrix | List of critical technologies | (UNIDO, 2005) |
| Technology planning approach (decision matrix) | Important technologies, ability, and attractiveness | Ability – attractiveness Matrix | Four types of TS in each important technology | (Tassel, 1983; Morin, 1985; Vernet et al, 1999 cited by Ghazinoory et al, 2009) |
| Technological SWOT analysis | S,W,O,T | SWOT Matrix | TSs | (Cardullo, 1996) |
| National nanotechnology strategy in Iran | Country's technological vision and goals, list of areas of technologies, STEEP, attractiveness, and capability | SWOT matrix, capability – attractiveness matrix | TSs | (Ghazinoory et al, 2009) |
| Technology sourcing decision | Cost of technology development, strategic impact of technology | Matrix presented by AD Little based on the strategic impact of technology and the anticipated cost of technology development. | Three technology sourcing decisions | (Floyd, 1998) |
| Booz-Allen & Hamilton methodology | List of areas of technologies, technology importance, position of technology, business attractiveness, and position of business | 2 matrixes: technology and business | Investment prioritization | (Pappas, 1984) |
| Chiesa's model | List of technologies and their applications | Technology – application matrix | five TSs, three main decisions (selection, acquisition, and timing) | (Chiesa, 1996; 2001) |
| Burgelman et al's model | Forces that shape TS | Model of forces that shape TS | three main decisions (selection, acquisition and timing) | (Burgelman, 2001) |

*continued on following page*

*Table 1. Continued*

| Method | Input | Process | Output | Reference |
|---|---|---|---|---|
| A. D. Little's model | list of the technologies, competitive impact of the technologies, and the firms' technological strength | Matrix of competitive impact of the technologies -the firms' technological strength | TSs | (Arasti et al, 2009) |
| Porter's model | list of the technologies, evaluation of the firm's capabilities and the required investments for technology development | Five forces model and value chain model | Two main decisions (selection and timing) | (Porter, 1985) |
| Hax & Majluf's model | Firm and/or business strategy, technology assessment, and internal scrutiny | The model of TS formulation | three main decisions (selection, acquisition and timing) | (Hax and Majluf, 1996) |
| McKinsey's Model | prospects for increased productivity and prospects for increased yield | Matrix of prospects for increased productivity and prospects for increased yield | Three main strategy | (Arasti et al, 2009) |

Ebrahimi et. al., 2013.

## MULTI AGENT SYSTEM (MAS)

The complexity of information systems (IS) does not stop growing and their development becomes increasingly complex, expensive, and difficult as well. The use of new advanced systems engineering techniques is essential to contribute to control the complexity of the encountered problems and to create required flexibility and adaptability. To provide the improvement of IS engineering, IS development phase based on multi-agent systems (MASs) is proposed. Different works tried to bring closer the MAS modelling and IS modelling, in different perspectives (Fazziki et al, 2010).

Decision-making processes in strategic planning are usually very complex and are frequently partitioned into sub-problems. In this case, the development of MASs provides a new way to solve complex problems by dividing them into a number of agents which cooperate to suggest strategies using their own beliefs, desires (or goals), and intentions. Strategy making is compounded by the ill-structured nature of the decision-making process. Thus, these systems extend the limit of bounded rationality by promoting improved understanding, better insights, and more extensive analysis.

In recent years, academicians and researchers have focused on using ISs in supporting strategy planning process. In the case of TS planning, there is a need of MASs particularly in conditions of SMEs. There are differences in the ways SMEs and large firms have attempted to practice toward the objective of improving their competitiveness. Besides, in a research such as Li and Li (2010), multiple agents are used with simulation for strategy planning to obtain an unbiased and consistent point estimator for relevant factors influencing the strategy planning.

According to Taghezout et al. (2009), in order to develop agent-based systems by using agent paradigm, there are various existing application domains such as software technology, robotics, and complex systems. Multi-agent decision systems and multi-agent simulation systems are recognized as two main MAS paradigm. One of the prominent characteristics of multi-agent decision system is that agents can participate in the system to make joint and participative decision as a group.

Strategic planning is recognized as a collaborative process of data collection, data analysis and decision making considering diverse factors influencing directions to achieve goals and objectives which requires involving of decision makers distributed across the company. It is therefore critical to facilitate their strategic planning by a distributed MAS. Development of TS in organizations will be more effective if it is seen mainly as a social rather than analytical process. Therefore designing a distributed support system which facilitates collaboration among decision makers can play an important role in facilitating the negotiation of strategy.

MASs as multi-agent decision systems are software systems have capacity for several autonomous software agents carrying out in a distributed environment. There are global objectives which are established through committing all or some group of agents to the termination of their conducts besides the local goals of each agent (Taghezout et al., 2009).

This approach has some advantages as the followings (Taghezout et al., 2009):

- It is recognized as a natural way for the purpose of controlling the complexity of large and highly distributed systems;
- It makes the construction of scalable systems easier since it is possible to add more agents in an easy way;
- Potentially, MASs are more robust and fault-tolerant than centralized systems.

An intelligent agent performs, reactively and/or proactively, interactive tasks tailored to a user's needs without humans or other agents telling it what to do. Agents are the result of a paradigm shift in developing application software. Software is no longer regarded as a tool. Rather, it is considered as an autonomous assistant to the users (Bui and Lee, 1999).

Agents have the following properties (Potiron et al., 2013):

- **Autonomy:** An agent encompasses individually goals, resources and competences; as such it performs its operation without human intermediary or any other intervention, and has a degree of control over its behaviors, actions, and internal condition. Agent adaptability means that an agent has the ability to control its own state and therefore can govern its own operation without any assistance and supervision from outside. This is one of the foremost consequences of agent autonomy.
- **Sociability:** The feature of an agent pertinent to having the ability to interact with other agents and possibly humans, through some kind of agent communication language is defined as sociability. This means that an agent is able to provide and ask for services.
- **Reactivity:** An agent, like this single agent that is described in this study, perceives and acts, to some degree, on its close environment; it can react in a timely manner to assumed changes that happen around it.
- **Pro-Activeness:** Beside some reactive agents that will simply act in response to stimulations from their environment, an agent may be able to exhibit goal-directed behavior by taking the initiative.

There are some techniques and mechanisms which are used for development of strategies and TSs as the followings:

- A distributed decision support system for strategic planning (Pinson et al, 1997): This study presents a general framework for the aim of creating a distributed strategic decision support system which can integrate advancements in distributed decision making and distributed artificial intelligence. This distributed system encompasses two specific characteristics. One of the features is that the users are able to intervene as human agents in the solution formation, and another one refers to distribution of strategic knowledge and domain knowledge in different agents which communicate through various blackboards and message passing. In this study, a case study is provided in the field of strategic marketing which describes the way of system's operations.
- Internet-based multi-agent intelligent system for developing marketing strategies (Li, 2007): This study illustrates the development and experimental evaluation of an internet-enabled multi-agent prototype system, named as Agent-Stra, for the purpose of developing marketing strategies, competitive strategies and associated IT/IS/e-commerce strategies.

- Multi-agent based hybrid intelligent systems for international marketing planning (Li & Li, 2010): An increase in complexity and competition in the global marketing causes new challenges for decision makers. In this study, the features of global marketing planning are discussed. The advantages and disadvantages of those methods and tools that may be used in the planning problem are analyzed. Then, a multi-agent-based hybrid intelligent framework for the aim of international marketing planning and associated internet strategy formulation is established. Additionally, a software prototype of the hybrid framework, called AgentsInternational, is created and presented, with initial evaluation results reported.
- Fuzzy inference system for technological strategy (Yu & Lo, 2009): This study presents a TS planning method that integrates patent analysis techniques with a fuzzy inference system (FIS). The proposed method uses global patent databases as sources of knowledge for strategy planning. The patent analysis techniques are applied to evaluate the technological competitiveness of a firm in a specific technology domain. A FIS for TS planning is developed. A case study related to construction industry is conducted. Two construction firms are selected for case study to illustrate the applicability of the proposed method based on different scenarios. The results show that the proposed method provides a useful and systematic approach for construction firm to plan effective TS.

# METHODOLOGY

## Winikoff Three-Layer Architecture

As stated by Henderson-Sellers and Giorgini (2005), one famous agent architecture is the 'beliefs', 'desires', and 'intentions' (BDI) architecture of Rao and Georgeff (1995). Padgham et al. (2008) summarizes this architecture in terms of three abstraction layers called philosophical/psychological, theoretical, and implementation. In order to formalize the internal architecture of a complex agent like the agent described in the present study, BDI paradigm is a classic paradigm. This paradigm makes it possible to design expressive and realistic agents, yet this paradigm is rarely applied in social simulations. This happens because most agent architectures which are established on the basis of the BDI paradigm are too complex to be understood and used by non-computer scientists. Furthermore, they are often very time-consuming in the case of computation and therefore not adapted to simulations with thousands of agents (Caillou et al., 2015).

'Beliefs', 'desires', and 'intentions' are seen as high-level, abstract, and externally ascribed features. 'Beliefs' represent the information available to an agent (e.g., about the environment or other agents). 'Desires' are regarded as the motivational state of the agent. They refer to objectives or situations that the agent would like to accomplish or bring about. 'Intentions' mean the deliberative state of the agent – what the agent has chosen to do. 'Intentions are desires to which the agent has to some extent committed.

As it can be seen in Table 2 and Figure 1, in psychology phase, answers of the questions where and how should be found which can be determined by 'belief', 'desire', and 'intention'. In design/model phase, a general map or 'world model' using design tools like Prometheus is designed and displayed. Additionally, 'goal' and 'goal commitment' are detected by defining procedures and 'action'. At last, in the implementation phase, using programming tools, knowledge base of the system is implemented and afterwards, system can be executed.

*Table 2. Relationship between terminologies of BDI model*

| Viewpoint | | Column 3 | Column 3+ Commitment | |
|---|---|---|---|---|
| Psychology | Belief | Desire | Intention | Wherewithal ('how') |
| Design/Model | World Model | Goal | Commitment | Plan |
| Implementation | Knowledge Base | - | - | Running (or instantiated Plan) |

Henderson-Sellers and Giorgini, 2005.

*Figure 1. Three layer-architecture of the system*

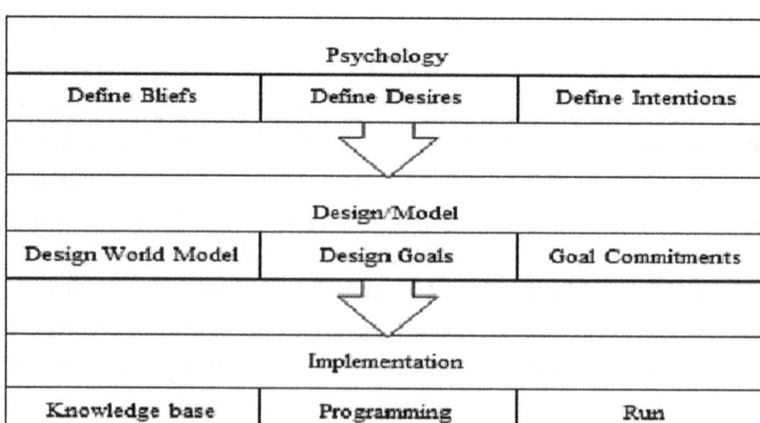

## TECHNOLOGICAL SWOT FACTORS

Main concern in using SWOT analysis is excessive simplification. Its quality is also dependent on the team which creates it. In order to find SWOT factors in this study, the author used interview and survey methods. Interview method is applied to find certain information and acquire definite factors by asking from managers and experts in renewable energy SMEs. Subsequently, in survey method by creating a questionnaire made by defined questions from the interview and sending it to forty selected experts, answers and factors were determined.

According to Table 3, internal factors represent strengths –those competences which offer a firm, competitive advantages compared with some similar companies, and weaknesses –those characteristics which generate competitive disadvantages. Every company presents a specific combination of strengths and weaknesses.

As it can be seen from Table 4, external factors represent opportunities –a combination of external elements, which give industry significant advantages, in conditions of a certain course it acts, and threats –a combination of external elements, which provoke some significant damage to industry in conditions of keeping the course of its action, a course that exists when these elements appear.

*Table 3. Internal factors*

| | |
|---|---|
| **IF1** | The quality of technologies provided by the company, |
| **IF2** | Access to required experience, knowledge, and information in the field of technology, |
| **IF3** | Reputation among customers in the technology field, |
| **IF4** | The ability of the company in the reverse engineering of the technology, |
| **IF5** | Capacity of idea generation and innovation in the technology field, |
| **IF6** | Use of well-known brand in the technology field, |
| **IF7** | Having the capacity of installing the technologies, |
| **IF8** | Having the capacity of maintenance and guarantee, |
| **IF9** | The diversity of technologies provided to the customers, |
| **IF10** | Having access to required skilled labor in the field of technologies, |
| **IF11** | Access to proper distribution networks in the field of technologies, |
| **IF12** | Capacity of financing from various sources, |
| **IF13** | Having access to the necessary resources in the technology field, |
| **IF14** | Efficient technology management, |
| **IF15** | The number of patents, |
| **IF16** | Capacity of performing the main processes of technology manufacturing. |

*Table 4. External factors*

| | |
|---|---|
| **EF1** | The existence of government policies on new energy boost, |
| **EF2** | The existence of potential customers and needs to be accountable, |
| **EF3** | The possibility of cooperation with research centers with the aim of accelerating the access to technologies, |
| **EF4** | Technologies fit with the national technological priorities, |
| **EF5** | The existence of competitive market. |
| **EF6** | Stability in customer needs and preferences which makes distance to provided technologies, |
| **EF7** | Lack of emergence of more appropriate alternative technologies, |
| **EF8** | The nonexistence of trade barriers and constraints, |
| **EF9** | The existence of international markets for the supply of technologies, |
| **EF10** | The country's economic stability, |
| **EF11** | Political stability of the region and country, |
| **EF12** | Adaptation of the use of new energy technologies with the current state of political and legislative (absence of lobbying for fossil fuels) |
| **EF13** | Public perception about the use of technology, |
| **EF14** | The nonexistence of strong competitors and technology providers. |

## TURNING TECHNOLOGICAL SWOT ANALYSIS INTO A TECHNOLOGY STRATEGIC PLAN

By considering SWOT analysis of Cardullo (1996) and gathered information, it is possible to develop a technological strategic plan based on Figure 2.

According to Cardullo (1996), 'SO' is an aggressive strategy, 'WO' means a turnaround oriented strategy, 'ST' is a diversification strategy, and 'WT' means a defensive strategy. For more clarification:

*Figure 2. The model of general technological strategic planning*

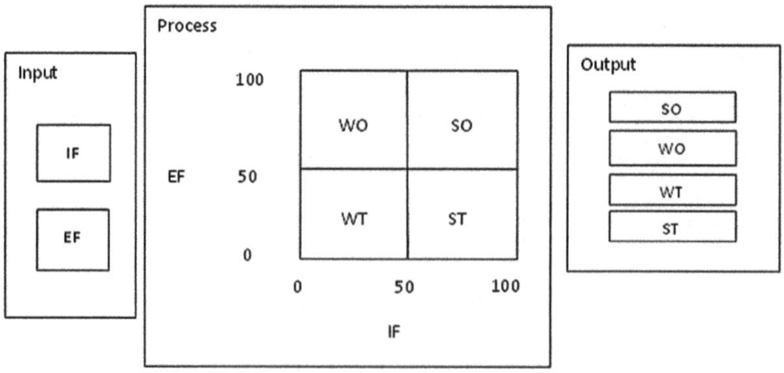

- **'SO':** If (IF =>50) and (EF =>50) then 'using IF and EF, maximize returns by taking a relatively higher degree of risk to have higher market share.'
- **'WO':** If (IF < 50) and (EF =>50) then 'solve the root cause of the low IF and by defining the crucial reasons behind it, restructure the company to improve the IF in the field of technologies.'
- **'ST':** If (IF =>50) and (EF < 50) then 'using IF, diversify your activities by expanding into markets or products that are related or unrelated to your current business.'
- **'WT':** (IF < 50) and (EF < 50) then 'cut the price of the technologies, add incentives or discounts to encourage customers to buy from you, and increase your advertising and marketing campaigns.'

## Phase 1: BDI Belief, Desire, Intention

In this phase, 'belief', 'desire', and 'intention' are defined. In the present research, there is one agent named as general technology strategy agent according to SWOT analysis. As it can be seen from Table 5, the 'belief', 'desire', and 'intention' of this agent are determined.

## Phase 2: Design With Prometheus Design Tool

The Prometheus Design Tool (PDT) is a graphical editor which supports the design tasks specified for designing agent systems. It is intended to support software engineers and developers in both developing and documenting the various aspects of specifying and designing a system using an agent oriented approach. The PDT allows a user to enter and edit a design, in terms of Prometheus concepts; check the design for a range of possible inconsistencies; and automatically generate a design report including diagrams. So, PDT supports the Prometheus methodology in a number of ways.

Figure 3 shows the design model of the agent with descriptions presented in Table 6.

*Table 5. BDI in the system*

| Intentions | Desires | Beliefs | Agent |
|---|---|---|---|
| According to 'IF', and 'EF', agent commits to deliver strategies in four mentioned categories. | This agent has following desires: 'SO', 'WO', 'ST', and 'WT'. | IF and EF | General Technology Strategy Agent |

*Figure 3. System design in Prometheus*

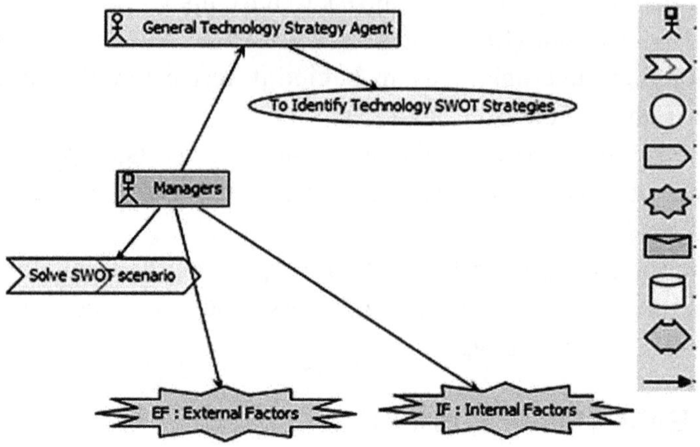


*Table 6. Agent, goal, actor, percept, scenario*

| Agents | Description |
|---|---|
| General Technology Strategy Agent | This is an agent which can propose general technology strategies according to the SWOT analysis in renewable energy SMEs. |
| **Goals** | **Description** |
| To identify 'Technology SWOT Strategies' | 'SO', 'WO', 'ST', and 'WT' |
| **Actors** | **Description** |
| Managers | All the managers in the organizational hierarchy |
| **Percepts** | **Description** |
| 'IF' | Table 2 |
| 'EF' | Table 3 |
| **Scenarios** | **Description** |
| Solve SWOT scenario | According to 'IF', and 'EF', agent proposes strategies in four mentioned categories |

## Phase 3: Implementation Phase

WIN-PROLOG is the leading Prolog compiler system for Windows-based PCs. Prolog is an established and powerful artificial intelligence (AI) language which provides a high-level and productive environment based on logical inference. It includes support for different types of rule-based inference. Figure 4 shows the system framework.

*Figure 4. System framework*

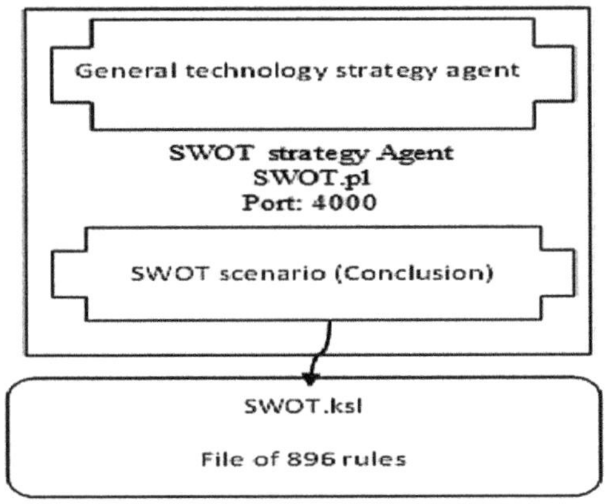


## SAMPLE CODES AND RESULTS

### Sample Codes of SWOT.ksl File

Some sample codes of SWOT.ksl considering 'the quality of technologies provided by the company' as an internal factor and 'the existence of government policies on new energy boost' as an external factor are presented as the followings:

```
relation continue1(Conclusion) if
rel_code_1 and
[
check(if1, >=, 50) and
check(ef1, >=, 50)
] and
Conclusion = 'Because of the quality of technologies provided
by the company is high and the existence of government policies
on new energy boost is high then we advise:using the quality of
technologies, and the existing government policies, maximize
returns by taking a relatively higher degree of risk to have
higher market share.' .
%%%%%%%%%%%%%%%%%%%%%%%%%%%%%%%%%%%%%%%%%%%%%%%%%%%%
relation continue1(Conclusion) if
rel_code_1 and
[
check(if1, <, 50) and
check(ef1, >=, 50)
```

```
] and
Conclusion = 'Because of the quality of technologies provided
by the company is low and the existence of government policies
on new energy boost is high then we advise:solve the root cause
of the low quality and by defining the crucial reasons behind
it, restructure the company to improve the quality of the
technologies' .
%%%%%%%%%%%%%%%%%%%%%%%%%%%%%%%%%%%%%%%%%%%%%%%%%%%%
relation continue1(Conclusion) if
rel_code_1 and
[
check(if1, >=, 50) and
check(ef1, <, 50)
] and
Conclusion = 'Because of the quality of technologies provided
by the company is high and the existence of government policies
on new energy boost is low then we advise:using the quality
of technologies, diversify your activities by expanding into
markets or products that are related or unrelated to your
current business.' .
%%%%%%%%%%%%%%%%%%%%%%%%%%%%%%%%%%%%%%%%%%%%%%%%%%%%
relation continue1(Conclusion) if
rel_code_1 and
[
check(if1, <, 50) and
check(ef1, <, 50)
] and
Conclusion = 'Because of the quality of technologies provided
by the company is low and the existence of government policies
on new energy boost is low then we advise:cut the price of
the technologies, add incentives or discounts to encourage
customers to buy from you, and increase your advertising and
marketing campaigns.' .
%%%%%%%%%%%%%%%%%%%%%%%%%%%%%%%%%%%%%%%%%%%%%%%%%%%%
relation swotstrategy(Conclusion) if
 q_if1(Conclusion) .
relation q_if1(Conclusion) if
 the answer to if1 is _ and
 q_if2(Conclusion) .
relation q_if16(Conclusion) if
```

```
 the answer to if16 is _ and
 q_ef1(Conclusion) .
relation q_ef1(Conclusion) if
 the answer to ef1 is _ and
 q_ef2(Conclusion) .
question if1
 'Enter a number between 0 and 100 for the quality of
technologies provided by the company ' ;
 input number
 because we want to know the quality of your products as an
internal factor .
question ef1
 'Enter a number between 0 and 100 for the existence of
government policies on new energy boost ' ;
 input number
 because we want to know the government policies as an external
factor.
```

## Other Sample Codes of SWOT.ksl

Some other sample codes of SWOT.ksl are presented in appendices:

- **Appendix 1:** Codes related to 'access to required experience, knowledge, and information in the field of technology' as an internal factor and 'public perception about the use of technology' as an external factor,
- **Appendix 2:** Codes related to 'having the capacity of installing the technologies' as an internal factor and 'the country's economic stability' as an external factor,
- **Appendix 3:** Codes related to 'efficient technology management' as an internal factor and 'technologies fit with the national technological priorities' as an external factor.

## Sample Code of SWOT.pl File

Some sample codes of SWOT.pl are presented in the section.

```
swotstrategy_agent:-
 agent_create(overall, handler, 4000),
 write(`swot agent Created!~M~J`),
```

```
write(`==================~M~J~M~J`),
write(`swot strategy agent is running at Port: `),
write(` 4000`),
write(`~M~J~M~J`).
```

## Sample Result of the System

The result of the system considering 'the quality of technologies provided by the company' as an internal factor and 'the existence of government policies on new energy boost' as an external factor in a sample case study is as the following:

`conclusion=``'Because of the quality of technologies provided by the company is high and the existence of government policies on new energy boost is high then we advise:using the quality of technologies, and the existing government policies, maximize returns by taking a relatively higher degree of risk to have higher market share.' .

The result of the system considering 'access to required experience, knowledge, and information in the field of technology' as an internal factor and 'Public perception about the use of technology' as an external factor in the same case study is:

`conclusion=``'Because of access to required experience, knowledge, and information in the field of technology is low and public perception about the use of technology is low then we advise: cut the price of the technologies, add incentives or discounts to encourage customers to buy from you, and increase your advertising and marketing campaigns.' .

The result of the system considering 'having the capacity of installing the technologies' as an internal factor and 'the country's economic stability' as an external factor in the same case study is:

Conclusion = 'Because of having the capacity of installing the technologies is high and the country's economic stability is low then we advise:use of the capacity of installing the technologies, diversify your activities by expanding into markets or products that are related or unrelated to your current business.' .

The result of the system considering 'efficient technology management,' as an internal factor and 'technologies fit with the national technological priorities' as an external factor in the same case study is:

Conclusion = 'Because of efficient technology management is low and technologies fit with the national technological priorities is high then we advise: solve the root cause of the low efficient technology management and by defining the crucial reasons behind it, restructure the company to improve the efficient technology management ' .

## CONCLUSION

In this study, development of the BDI agent – based system for suggesting general TSs in renewable energy SMEs was illustrated. This system was designed on the basis of the technological SWOT-analysis. SWOT analysis is a precursor to strategic planning and is performed by forty experts who can assess the company from technological perspective. Identification of SWOT factors including internal strengths and weaknesses, as well as external opportunities and threats took place by applying interview and survey methods. The TSs were identified in four categories: aggressive strategy, turnaround oriented strategy, diversification strategy, and defensive strategy. Such analyses promote a better understanding of the companies and their environment, and knowing the best way to achieve the goals.

Afterwards, according to the three-layer architecture cited by Winikoff, system architecture was described concerning layers of psychological, theoretical, and implementation. Subsequently, the system was executed and used by experts and managers in renewable SMEs and they validated the results. Through validation phase, experts admit that TS planning due to the turbulent environment of the industry is complicated and using a system like this improves the quality of TS planning.

## REFERENCES

Arasti, M. R., & Packniat, M. (2011). *A Classification of Models for Technology Strategy Formulation*. Retrieved October 24, 2014, from http://www.nrisp.ac.ir/jstp/browse.php?a_code=A-10-28-64&sid=1&slc_lang=en

Bui, T., & Lee, A. N. D. (1999). An agent-based framework for building decision support systems. *Decision Support Systems*, *25*(3), 225–237. doi:10.1016/S0167-9236(99)00008-1

Caillou, P., Gaudou, B., Grignard, A., Truong, C. Q., & Taillandier, P. (2015). *A simple-to-use BDI architecture for agent-based modeling and simulation.* The Eleventh Conference of the European Social Simulation Association (ESSA 2015), Groningen, Netherlands. Retrieved August 29, 2017, from https://hal-univ-tlse3.archives-ouvertes.fr/hal-01216165/document

Cardullo, M. W. (1996). *Introduction to managing technology*. New York: John Wiley & Sons.

Chiesa, V. (1996). Towards a new framework for technology strategy formulation. *Proceedings of the 3rd International Product Development Conference,* 165-178.

Chiesa, V. (2001). *R&D Strategy and Organization*. London: Imperial College Press. doi:10.1142/p223

Ebrahimi, M., Moghbel Baerz, A., Khodadad Hosseini, S. H., & Azar, A. (2013). A New Model of Petrochemical Technology Strategic Planning. *International Journal of Business Administration*, *4*(2), 57–73. doi:10.5430/ijba.v4n2p57

Edvardsson, I. R., & Durst, S. (2013). The Benefits of Knowledge Management in Small and Medium-sized Enterprises. *Procedia: Social and Behavioral Sciences*, *81*(28), 351–354. doi:10.1016/j.sbspro.2013.06.441

Fazziki, A. E., Nouzri, S., & Sadgal, M. (2010). Towards an MDA Based Multi-agent Approach for Information System Development. *International Journal of Digital Society*, *1*(3), 190–198. doi:10.20533/ijds.2040.2570.2010.0024

Floyd, C. (1998). *Managing technology for corporate success*. Aldershot, UK: Gower.

Ghazinoory, S., Divsalar, A., & Soofi, A. S. (2009). A new definition and framework for the development of a national technology strategy: The case of nanotechnology for Iran. *Technological Forecasting and Social Change*, *76*(6), 835–848. doi:10.1016/j.techfore.2008.10.004

Henderson-Sellers, B., & Giorgini, P. (2005). *Agent-Oriented Methodologies*. Idea Group Pub. doi:10.4018/978-1-59140-581-8

Houben, G., Lenie, K., & Vanhoof, K. (1999). A Knowledge-based SWOT-analysis System as an Instrument for Strategic Planning in Small and Medium Sized Enterprises. *Decision Support Systems*, *26*(2), 125–135. doi:10.1016/S0167-9236(99)00024-X

Kubíčková, L., Votoupalová, M., & Toulová, M. (2014). Key Motives for Internationalization Process of Small and Medium–Sized Enterprises. *Procedia Economics and Finance*, *12*, 319–328. doi:10.1016/S2212-5671(14)00351-7

Li, S. (2007). AgentStra: An Internet-based multi-agent intelligent system for strategic decision-making. *Expert Systems with Applications*, *33*(3), 565–571. doi:10.1016/j.eswa.2006.05.018

Li, S., & Li, J. Z. (2010). AgentsInternational: Integration of multiple agents, simulation, knowledge bases and fuzzy logic for international marketing decision making. *Expert Systems with Applications*, *37*(3), 2580–2587. doi:10.1016/j.eswa.2009.08.022

Padgham, L., Thangarajah, J., & Winikoff, M. (2008). Prometheus design tool. *Proceedings of the Twenty-Third AAAI Conference on Artificial Intelligence*. Retrieved January 13, 2016, from https://www.aaai.org/Papers/AAAI/2008/AAAI08-330.pdf

Pappas, C. (1984). Strategic management of technology. *Journal of Product Innovation Management, 1*(1), 30–35. doi:10.1016/S0737-6782(84)80040-5

Pinson, D. S., Louq, J. A., & Moraitis, P. (1997). A distributed decision support system for strategic planning. *Decision Support Systems, 20*(1), 35–51. doi:10.1016/S0167-9236(96)00074-7

Potiron, K., El Fallah Seghrouchni, A., & Taillibert, P. (2013). *From fault classification to fault tolerance for multi-agent systems*. Springer. Retrieved August 28, 2017, from http://www.springer.com/gp/book/9781447150459

Shane, S. (2009). *Technology strategy for managers and entrepreneurs*. Pearson/Prentice Hall.

Taghezout, N., Adla, A., & Zaraté, P. (2009). A multi-agent framework for group decision support system: Application to a boiler combustion management system (GLZ). *International Journal of Software Engineering and Its Applications, 3*, 2, 9–20.

Yu, W.-D., & Lo, S.-S. (2009). Patent analysis-based fuzzy inference system for technological strategy planning. *Automation in Construction, 18*(6), 770–776. doi:10.1016/j.autcon.2009.03.003

## APPENDIX 1

```
relation continue1(Conclusion) if
rel_code_1 and
[
check(if2, >=, 50) and
check(ef13, >=, 50)
] and
Conclusion = 'Because of access to required experience,
knowledge, and information is high and public perception about
the use of technology is high then we advise:using the access
to required experience, knowledge, and information, and public
perception about the use of technology, maximize returns by
taking a relatively higher degree of risk to have higher market
share.' .
%%%%%%%%%%%%%%%%%%%%%%%%%%%%%%%%%%%%%%%%%%%%%%%%%%%%%
relation continue1(Conclusion) if
rel_code_1 and
[
check(if2, <, 50) and
check(ef13, >=, 50)
] and
Conclusion = 'Because of access to required experience,
knowledge, and information is low and public perception about
the use of technology is high then we advise:solve the root
cause of the low access to required experience, knowledge,
and information and by defining the crucial reasons behind
it, restructure the company to improve the access to required
experience, knowledge, and information ' .
%%%%%%%%%%%%%%%%%%%%%%%%%%%%%%%%%%%%%%%%%%%%%%%%%%%%%
relation continue1(Conclusion) if
rel_code_1 and
[
check(if2, >=, 50) and
check(ef13, <, 50)
] and
Conclusion = 'Because of the access to required experience,
knowledge, and information is high and public perception about
the use of technology is low then we advise:using the access
```

```
to required experience, knowledge, and information, diversify
your activities by expanding into markets or products that are
related or unrelated to your current business.' .
%%%%%%%%%%%%%%%%%%%%%%%%%%%%%%%%%%%%%%%%%%%%%%%%%%%%
relation continue1(Conclusion) if
rel_code_1 and
[
check(if2, <, 50) and
check(ef13, <, 50)
] and
Conclusion = 'Because of access to required experience,
knowledge, and information is low and public perception about
the use of technology is low then we advise:cut the price of
the technologies, add incentives or discounts to encourage
customers to buy from you, and increase your advertising and
marketing campaigns.' .
%%%%%%%%%%%%%%%%%%%%%%%%%%%%%%%%%%%%%%%%%%%%%%%%%%%%
relation q_if2(Conclusion) if
 the answer to if2 is _ and
 q_if3(Conclusion) .
relation q_ef12(Conclusion) if
 the answer to ef12 is _ and
 q_ef13(Conclusion) .
relation q_ef13(Conclusion) if
 the answer to ef13 is _ and
 q_ef14(Conclusion) .
question if2
 'Enter a number between 0 and 100 for the access to required
experience, knowledge, and information ' ;
 input number
 because we want to know the access to required experience,
knowledge, and information as an internal factor .
question ef13
 'Enter a number between 0 and 100 for the public perception
about the use of technology ' ;
 input number
 because we want to know the public perception about the use of
technology as an external factor .
```

## APPENDIX 2

```
relation continue1(Conclusion) if
rel_code_1 and
[
check(if7, >=, 50) and
check(ef10, >=, 50)
] and
Conclusion = 'Because of having the capacity of installing
the technologies is high and the country's economic stability
is high then we advise:using the capacity of installing the
technologies, and the country's economic stability, maximize
returns by taking a relatively higher degree of risk to have
higher market share.' .
%%%%%%%%%%%%%%%%%%%%%%%%%%%%%%%%%%%%%%%%%%%%%%%%%%%%
relation continue1(Conclusion) if
rel_code_1 and
[
check(if7, <, 50) and
check(ef10, >=, 50)
] and
Conclusion = 'Because of having the capacity of installing the
technologies is low and the country's economic stability is
high then we advise:solve the root cause of the low capacity of
installing the technologies and by defining the crucial reasons
behind it, restructure the company to improve the capacity of
installing the technologies ' .
%%%%%%%%%%%%%%%%%%%%%%%%%%%%%%%%%%%%%%%%%%%%%%%%%%%%
relation continue1(Conclusion) if
rel_code_1 and
[
check(if7, >=, 50) and
check(ef10, <, 50)
] and
Conclusion = 'Because of the capacity of installing the
technologies is high and the country's economic stability
is low then we advise: using the capacity of installing the
technologies, diversify your activities by expanding into
markets or products that are related or unrelated to your
current business.' .
```

```
%%%%%%%%%%%%%%%%%%%%%%%%%%%%%%%%%%%%%%%%%%%%%%%%%%%%
relation continue1(Conclusion) if
rel_code_1 and
[
check(if7, <, 50) and
check(ef10, <, 50)
] and
Conclusion = 'Because of the capacity of installing the
technologies is low and the country's economic stability is
low then we advise: cut the price of the technologies, add
incentives or discounts to encourage customers to buy from you,
and increase your advertising and marketing campaigns.' .
%%%%%%%%%%%%%%%%%%%%%%%%%%%%%%%%%%%%%%%%%%%%%%%%%%%%
relation q_if6(Conclusion) if
 the answer to if6 is _ and
 q_if7(Conclusion) .
relation q_if7(Conclusion) if
 the answer to if7 is _ and
 q_if8(Conclusion) .
relation q_ef9(Conclusion) if
 the answer to ef9 is _ and
 q_ef10(Conclusion) .
relation q_ef10(Conclusion) if
 the answer to ef10 is _ and
 q_ef11(Conclusion) .
question if7
 'Enter a number between 0 and 100 for having the capacity of
installing the technologies ' ;
 input number
 because we want to know the capacity of installing the
technologies as an internal factor .
question ef10
 'Enter a number between 0 and 100 for the country's economic
stability ' ;
 input number
 because we want to know the country's economic stability as an
external factor .
```

# APPENDIX 3

```
relation continue1(Conclusion) if
rel_code_1 and
[
check(if14, >=, 50) and
check(ef4, >=, 50)
] and
Conclusion = 'Because of efficient technology management is
high and technologies fit with the national technological
priorities is high then we advise:using the efficient
technology management, and technologies fit with the national
technological priorities, maximize returns by taking a
relatively higher degree of risk to have higher market share.'
.
%%%%%%%%%%%%%%%%%%%%%%%%%%%%%%%%%%%%%%%%%%%%%%%%%%%%%%%%%%%%%
relation continue1(Conclusion) if
rel_code_1 and
[
check(if14, <, 50) and
check(ef4, >=, 50)
] and
Conclusion = 'Because of efficient technology management is
low and technologies fit with the national technological
priorities is high then we advise: solve the root cause of the
low efficient technology management and by defining the crucial
reasons behind it, restructure the company to improve efficient
technology management ' .
%%%%%%%%%%%%%%%%%%%%%%%%%%%%%%%%%%%%%%%%%%%%%%%%%%%%%%%%%%%%%%%
relation continue1(Conclusion) if
rel_code_1 and
[
check(if14, >=, 50) and
check(ef4, <, 50)
] and
Conclusion = 'Because of efficient technology management is
high and technologies fit with the national technological
priorities is low then we advise: using the efficient
technology management, diversify your activities by expanding
```

```
into markets or products that are related or unrelated to your
current business.' .
%%%%%%%%%%%%%%%%%%%%%%%%%%%%%%%%%%%%%%%%%%%%%%%%%%%%
relation continue1(Conclusion) if
rel_code_1 and
[
check(if14, <, 50) and
check(ef4, <, 50)
] and
Conclusion = 'Because of efficient technology management is low
and technologies fit with the national technological priorities
is low then we advise: cut the price of the technologies, add
incentives or discounts to encourage customers to buy from you,
and increase your advertising and marketing campaigns.' .
%%%%%%%%%%%%%%%%%%%%%%%%%%%%%%%%%%%%%%%%%%%%%%%%%%%%
relation q_if13(Conclusion) if
 the answer to if13 is _ and
 q_if14(Conclusion) .
relation q_if14(Conclusion) if
 the answer to if14 is _ and
 q_if15(Conclusion) .
relation q_ef3(Conclusion) if
 the answer to ef3 is _ and
 q_ef4(Conclusion) .
relation q_ef4(Conclusion) if
 the answer to ef4 is _ and
 q_ef5(Conclusion) .
question if14
 'Enter a number between 0 and 100 for the efficient technology
management, knowledge, and information ' ;
 input number
 because we want to know the efficient technology management as
an internal factor .
question ef4
 'Enter a number between 0 and 100 for the technologies fit
with the national technological priorities ' ;
 input number
 because we want to know the technologies fit with the national
technological priorities as an external factor .
```

# Chapter 3
# Scenario Planning for Technical Knowledge Development

**Maryam Ebrahimi**
*Azad University, Iran*

## ABSTRACT

*Localizing and creating technical knowledge in the developing countries to make them able to compete in the international arena will not be an easy task to undertake as it involves complexities and uncertainties. In this regard, methods of modeling and simulation help understanding the reality and thereby delineating the future. This paper discusses using system dynamics methodology in the Petrochemical Research and Technology Company (Iran). After determining causal loop diagrams, a dynamic model, and validation test, in order to improve the system, related scenarios to the research are proposed. The results show that research financing is a major obstacle in the course of growth and development; accordingly, some scenarios are recommended to increase technical knowledge through raising the initiation and completion rates of research projects. The paper closes with a description of the suggested scenarios.*

## INTRODUCTION

Increasingly assumed to be important in every aspect of business life, technology plays a significant role in making strategic decisions to be highly focused on global considerations (Lall, 1993). Meanwhile, industries working in the developing countries are not capable enough to compete with ever evolving modern technologies

DOI: 10.4018/978-1-5225-3226-2.ch003

(Putranto et al., 2003). Moreover, how well firms in the developing countries can take part in the global market depends on their abilities in terms of implementing new technologies and how well they are in reinforcing their knowledge-based competitive advantages (Wang, 1997).

Leonard-Barton (1995) presented a continuum in which the appearance of technological dexterity was seen to be extended from equipment purchase by an acquirer to complete technology transfer. As the first two levels, assembly or turnkey operations and also component adaptation and localization are presumably undertook within the developing countries. According to Marcus (1992), although the owner may not tend to reveal all about a technology, the acquirer attempts to capture as much knowledge as possible, so as to decrease its dependency on the supplier. However, one may note that buying a new machine or implementing a turnkey project does not necessarily provide an acquirer with "a new technology" to be considered as a replacement for technology transfer (Leonard-Barton, 1995). Furthermore, as greater compliance with a set of procedures requires larger deal of know-how, automation does not necessarily result in more effective manufacturing when a technology adapted to local conditions is concerned (Bohn, 1994).

Nevertheless, as the developing countries are not going to remain a follower in their efforts to compete in the international arena, they have to implement their technology policies in terms of product redesign and independent design of products, as the third and fourth levels, respectively. It should be noted that gaining the first-mover advantage according to the redesigned or independently adapted technologies will not be easy for the developing countries. As such, the developing countries have tried to rapidly develop technologies within recent years, i.e. they attempt to develop technical knowledge. This makes them independent of foreign countries, able to generate more money and capital income. Moreover, it improves the quality of life for the people in these nations, and promotes industries; especially oil related ones which comprise a dominant portion of industries in several developing countries.

The present study focuses on technical knowledge development in an Iranian company, namely Petrochemical Research and Technology Company (here after referred to as NPC-RT), working within the research and development sector of the petrochemical industry. With a quick look at the activities done in the company, the importance of technical knowledge is clear and evident; it is, in general, the basis for commercialization process leading to profit generation in a company or industry. Here, technical knowledge refers to knowledge of the petrochemical and chemical synthesis and catalysts obtained by the use of library studies, laboratory investigations and bench scale. Along with this definition, a research system, which deals with multitude complexities and dynamic behavior, consists of all three functions.

In spite of the fact that research is realized and established almost in all organizations, sophisticated and rapidly changing information, knowledge, and technology is the foundation for a vast majority of products and services. Nevertheless, although technology is everywhere, its research and development are still faced with tremendous problems for research and development (R&D) organizations, technology developers and implementers. In this study, system dynamics (SD) methodology is applied to clarify and illustrate the complexities of research system in NPC-RT and after determining causal loop diagrams, a dynamic model, and validation test, in order to improve the system, related scenarios to the research are proposed.

## AN INTRODUCTION TO NPC-RT COMPANY

Creating the required technical knowledge is the main objective followed by NPC-RT. That is, the company is to decide about the pathways to enhance the technical knowledge. Hereby, the company should concentrate on expanding its research projects. A rise in research projects may occur as a consequence of two changes: initiating greater research projects while completing further research projects at the same time. The initiation of research projects is influenced by the time to perform research projects as well as the company's perceived research capacity. As the time to perform research projects lengthen, lower numbers of research projects are initiated. Additionally, the company's perceived research capacity is related to the deal of research staff.

Besides, the company requires a larger deal of such resources as research staffs, budget, and equipment to accomplish more research projects. For instance, one may expect research projects to be completed within the scheduled time provided adequate research staffs are available at the moment. In contrast, a lack of adequate staffs results in an increase in the time to perform research projects. Furthermore, as more research projects are performed, the research experience is enhanced, so that the higher number of research projects completed, the greater research experience will be created contributing to reduced time to perform further research projects.

## SD METHODOLOGY

SDM may contribute to the understanding of a system by extracting its working mechanisms' essential structures and developing efficient management scenarios based on an analysis of inherent feedback structures to the system (Park, 2014). Here the ontological principle is that systems are assumed to exist in the real world with its epistemological premise being essentially positivistic describing the underlying structure of the real world in terms of flows and rates (Rodriguez-Ulloa and Paucar-Caeres, 2004).

Once the problem area or management's policy issues of concern were identified, in order to use SD as a successful learning tool, SD models use positive and negative feedback loops, called casual loops, to identify the dynamics arising from these interactions and to reveal the structure of a system. As such, it is possible to express the system's behavior during a given interval (Zaim et al., 2013; Zhang et al., 2013). SD needs to be engage with mental models which, as the basis of decision-making process, hold the most significant information about social situations. Hence, actions undertaken to elicit, debate and facilitate change in the mental models by decision-makers can contribute to better management of a system (Lane, 2000). The main pitfall in the approach based on mental models is that their dynamic consequences cannot be mentally simulated to provide the primary rational for using SD modeling to support dynamic decision-making (Doyle and Ford, 1998).

Carried out in dynamic modeling are the development of the system diagram showing the main variables and issues involved in the system of interest, construction of computer simulation model, over-time simulations, generation and comparison of the model behavior with the historical trends, and model verification. Subsequently, policies and scenarios are postulated and tested.

Thanks to the recent significant advancements achieved in SD, which have made the subject more accessible to policy-makers, SD can be seen as the leader in the process of mapping policymakers' knowledge or mental model as well as other information about the business or social system, and then converting them into models and simulations (Morecroft, 1988). SD also fits into the strategy formulation process with three phases: analysis, planning, and control. In other words, SD supports the analysis phase of the strategy planning process by iteratively structuring the problem followed by testing and refining the problem structure. So, SD is applied for planning which is also an iterative process engaged with evaluation, selection and implementation of strategies. Forecasting expected performance, SD provides an essential element to the control phase any deviation from which indicates the need for additional analyses (Lyneis, 2009). SD is applied in various researches (Feng, 2012; Liuguo et al., 2012; Hoffmann et al., 2013; Rasjidin et al., 2012; Kusumaningdyah et al., 2013; Lee and Chung, 2012; Abada et al., 2013); however, to the best of our knowledge, there is few researches regarding technical knowledge development.

## MENTAL MODELING

Mental modeling deals with experiences which can influence one's ability to respond through his/her prior knowledge; it may also lead to build preconceptions on how to respond (Wilner and Birnberg, 1986). Besides, it acts as a filter through which an actor may see the problem (Kolkman et al., 2005). Mental modeling is considered to be the final result of perception, imagination, and the comprehension of discourse (Goldvarg and Johnson-Laird, 1983). Existing within the decision-maker before he/she even comes to the decision situation, these conditions can affect the way he/she understand the task instructions and the problem itself (Wilner and Birnberg, 1986).

According to Kolkman et al., 2005, a mental model lives at the heart of every decision-making process. There may be more than one mental model, or some sorts of collective model. Mental models have been used to explain behavior exhibited by industrial and economic systems (Forrester, 1991). Managerial mental models represent simplified knowledge structures or cognitive representations of how the business environment behaves (Gary, 2008). Some of the characteristics of mental models are as follows: 1) they are not always accurate (Gentner, 2002), 2) they are fuzzy, incomplete, and imprecise (Forrester, 1971), and 3) one cannot mentally simulate dynamic consequences (Doyle and Ford, 1998).

Influencing and being influenced by every step in the course of decision-making process, mental models determine what to be examined and what perspectives to be developed. Once perspectives were developed, the mental models are updated based on the insight gained. That is, something is learnt leading tacit knowledge to be created (Kolkman et al., 2005). Therefore, a mental model is to be constantly updated to keep the track of changes in circumstances (Li, 2011). In other words, a mental model needs learning to be changed (Johnson-Laird, 1983). For an organizational renewal to be realized, there is a need to such exhaustive adjustment phase in developing the firm's top managers' mental models once significant changes in the environment are witnessed (Barr, 1992). Besides, just like the managers, the academics have referred to organizational learning as the process wherein changes in shared understandings or shared mental models are realized (Senge and Sterman, 1992).

Using mental models as decision heuristics (as people frequently do) is based on the following assumptions (Glick et al., 2012):

- There exist mental models fitted to the purpose of taking action.
- Mental models are able to provide us with decision premises in those situations where we fail to determine what action to take.
- We will not confront with the components (or the possibility of examining the components or premises) of our mental models until the time when we take some action usually presented in the form of a decision.
- Decisions are taken within dynamic contexts.

## SCENARIO PLANNING

As a methodology encompassing a range of processes, scenario planning provides an opportunity to challenge our mental models against external landscape within which we operate, so as to promote informed and enhanced decision-making (Saliba, 2009). It is a well-organized method to imagine possible futures companies undertaken regarding a wide range of issues. Scenario planning draws a picture of the future developments of a given system in terms of interlinked qualitative and quantitative elements working toward promotion of internal consistency inside the system (Zahradníþková and Vacík, 2014). As an important supporting tool for governments, industries and communities, scenario planning helps them getting ready and planning for the future, managing risks and utilizing possible opportunities (Greiner et al., 2014). Companies can utilize scenario planning as a method to either generate or consider transformational changes (Tushman and Romanelli, 1985).

According to Buytendijk Et Al. (2010), scenario planning differs from other planning methods in a number of ways:

- Scenarios examine the combined effect of a variety of uncertainties standing side by side as equals.
- Scenarios alter multiple variables at a time while not necessarily keeping others unchanged.
- Going beyond objective analyses, scenarios encompass subjective interpretations.

Generally in scenario planning, learning refers to challenging those assumptions taken for granted about the organization as well as its environment (Chermack, 2005). Utilizing scenarios to adjust mental models in the course of strategic learning is an approach via which scenarios and scenario planning may provide new different insights into the world, such that one can share and challenge the knowledge about implicit processes and functions (Korte and Chermack, 2007). Scenario planning can be seen as a mean for companies by which to develop or improve their dynamic capabilities using tacit accumulation of experiences obtained from the hypothetical responses, knowledge representation in the course of simulations, and knowledge codification through the written representation of scenario planning decisions/results (Zollo & Winter, 2002). Scenario planning and SD have a much mutual basis. Mental models play the most important role in these methods. Using a misinformed mental model can lead scenario planning to develop an equally misinformed scenario.

## APPLICATION OF SD IN RELATED STUDIES

Research and development process is mostly about intangible and have to some extent a random probability pattern. It is generally perceived that the outcome of industrial R&D is irresolute and unclear. In many studies, researchers analogized research and development to a production function of material goods. For this kind of R&D processes, inputs are the allocated resources such as budget, number of human resources or laboratory equipment. The number of innovations or patents can be recognized as output (Milling and Maier, 2009).

In the related literature, there are many attempts to describe prominent aspects of R&D strategy, innovation process, and innovation diffusion. Kim and Choi (2009) discussed about technological innovation. According to their viewpoints, investment in activities relevant to R&D is the beginning of technological change. Having technological innovation as a result of R&D activities put the company in a more favorable position because it causes a better performance of technological productivity which in turn enhances its market share. And as a circular pattern, attained profit via the higher market power will be invested again for having more technological innovation and these processes will be repeated again. Therefore, technological innovation behaves according to a causal loop such as: "investment in R&D→accumulation of technological knowledge→product innovation →process innovation →new product & service production→profits maximization →reinvestment in R&D activities." It is required to have a comprehensive approach to analyze the numerous interactions between the different stages of the innovation process. Therefore, using models which consider these interactions should be considered. In the case of using appropriate models as tools, it is possible to have a better understanding of innovation processes, generate and evaluate strategies to improve the processes (Milling and Maier, 2009). One of the key factors for having improved performance of innovation process is the productivity of R&D activities as its first step.

In the case of production process, investment in R&D activities is one of the critical factors as it includes resources such as human and material resources in addition to information and knowledge. Besides including specific and particular techniques employed in the production process, R&D investment has scientific knowledge such as achieved experience and technical know-how. Technical know-how as a concept which covers or affects a large area that includes required knowledge and skill for accumulating, using, improving, changing, and building technologies. R&D investments make technological innovation possible because of its influence on increasing the R&D intensity and the accumulated R&D capability. However, many companies have not achieved success in having a better performance in R&D, despite their heavy investments in R&D.

According to Monga (2001), the process of a technology development includes diverse R&D activities. There is a lack of a system level understanding of the dynamics of the technology development process, both from management and process perspectives. This study focuses on a part of technology development which is R&D process for the purpose of technical knowledge development (technical know-how). In Table 1, literatures related to dynamic R&D processes and in Table 2, literatures related to dynamic technology innovation have been classified

## MAPPING MODELS WITH SDM IN THE COMPANY

In accordance with the specified goals of the system, the dynamic model developed in this chapter fits to the situations of research processes based on a two-year dataset and forecasting patterns for the coming three years by the managers and using modeling software called VENSIM PLE. The basis of the dynamic model is consistent with Figure 1 which shows the research mechanisms according to the SD theory. There are two stock variables in research sector. The first variable is current research projects referring to the pile of projects already started but not finished yet; i.e. current research projects variable is equivalent to the difference between research completion and initiation rates. The second variable, completed research projects, refers to the accumulation of completed research projects.

After identifying main stock and rate variables, one may continue with identifying the parameters affecting the rates. The completion rate of research is influenced by the time to perform the research project which is the main factor affecting the company's performance. Provided all research resources (including budget, equipment, and staff) are available with the staffs having enough deal of experience in terms of the current projects, then it is likely to go on a time schedule free of delay. Namely, lack of required experience and inadequate resources bring about delays in research. Here only the most important causalities for research are considered.

Figure 2 shows that the more current research projects, the larger deal of human resources are required for research. Provided adequate human resources for conducting the current research projects are available at the moment, the research projects are likely to complete within the expected time. In contrast, if the required research human resources exceed the current research human resources, there will be a lack of adequate research human resources which results in increasing the time to perform research project. Figure 3 shows that through performing research projects, the research experience will be enhanced. The higher completion rate of research projects creates larger deal of research related experience. By increasing the research experience, time to perform research project will be reduced, the completion rate of research projects will be enhanced and this mechanism strengthens itself.

*Table 1. Literature review related to dynamic R&D processes*

| Collection of models of R&D projects dynamics and resource allocation issues, | (Roberts,1978) |
|---|---|
| Effect of human resource allocation, sequential dependencies of and feedback between projects on workload of the R&D organization, | (Weil et al., 1973;1978) |
| Describing different programs such as technology programs, new product exploration programs and product development programs and focusing on crucial features such as the company's technology base, R&D activities and products in the market, | (Weil,1973) |
| Describing an issue of allocating resources to later stages of development instead of early stages which leads to poor performance of product development, | Repenning (2000; 2001) |
| Applying SD models to explore the relationships between R&D activity, pricing strategy and product diffusion, | Milling 1996; 2002; Milling and Maier 1993; 2001; Maier 1998) |
| Developing a simulation model to explain the role R&D can play in shaping the dynamics of a firm, | (Moizer and Towler, 2007) |
| Studying R&D portfolio strategy from the information perspective, | Gino and Pisano (2005; 2006) |
| • Presenting a general framework for R&D strategy including: (1) architecture, (2) processes, (3) people, and (4) portfolio. This framework focuses on resources selection and tactics to use those resources in a logical and consistent way,<br>• Discussing different decisions and choices affecting R&D performance such as: the size and location of R&D facilities, the division of labor between various groups, the choice of technologies used inside the R&D organization, the selection of personnel, the allocation of resources, the design of processes for managing projects, and other factors, | (Pisano, 2012) |
| Presenting policies to govern the acquisition of resources as R&D strategy issues, | (Roberts, 1978) |
| Presenting the links between resource acquisition and allocation and different aspects of technology strategy such as technical advancement of the company's products, the mastery of the technologies used in products, the nature of new technology programs, and the priorities for allocating resources to various activities, | (Weil, 2007) |
| Presenting the ability to manage R&D projects particularly in decisions such as project selection, resource allocation, project termination, and portfolio management and its huge impacts on competitive performance, | (Clark and Wheelwright, 1995) |
| Studying applications of SD contribute to the strategy field, | (Gary et al., 2008) |
| Applications of SD in innovation performance and innovation diffusion. | Milling, 1986; 1987; 1991; 1996; Milling and Maier, 1996; 2001; Rogers, 1983; Weil et al., 1987; Weil, 2007) |

Khodabakhshian et al., 2013.

*Table 2. Literature review related to dynamic technology innovation*

| | |
|---|---|
| Studies about management of technology innovation and focusing on industry dynamics | (Abernathy and Clark, 1985; Arthur, 1988; Carroll and Hannan, 2000; Geroski, 2001; Hall, 1994; Klepper, 1997; Pavitt, 1984; Tushman and Anderson, 1986; Walsh, 1984) |
| Studies about management of technology innovation and focusing on diffusion of technologies | (Bresnahan and Trajtenberg, 1995; David, 1990; Griliches, 1998; Kline & Rosenberg, 1986; Nelson and Phelps, 1966; Rogers, 1983) |
| Studies about organizational capacity or capability of technology innovation, focusing on R&D intensity and its effect on the level of organizational innovation and knowledge | (Godener and Soderquist, 2004; Boer, 1999; Jaffe et al., 2000; Griliches, 1998; Moizer and Towler, 2001; Bruce and Zander, 1992; Nonaka, 1994; Pisano, 1996; Tyre and Hippel, 1997; Van de Ven and Douglas, 1992) |
| Studies about innovation process, focusing on the level of organizational innovation and knowledge, and its effect on product innovation or process innovation | (Bean and Christopher, 1964; Carliss and Clark, 2000; Clark, 1985; Damanpour and Gopalakrishnan, 1997; Linard and Dvorsky, 2001; Marc and DeTore, 1999; Marc, 1996; Tertzakian, and Utterback, 1997; Repenning and Sterman,1998; Pisano, 1996; Cooper and Kleinschmidt,1998; Rosenberg & Kline,1986; Simon, 1981; Ulrich,1995; Utterback and Abernathy, 1978; Wheelwright and Clark, 1992; Stamboulis et el., 2000) |
| Studies about innovation process and focusing on independency of product and process innovation | (Ettlie and Ernesto, 1998; Milling and Stumpfe, 2000; Kotha and Orne, 1988; Kim et al, 1992; Ettlie, 1998; Stamboulis et al., 2002; Eppinger and Adler, 1999; Kotha and Orne, 1989) |
| Studies about innovation process, focusing on product or process innovation, and its effect on commercialization | (Patterson, 1985; Glöβler, 1996; Vivero, 1997; Maier, 1998; Kirkwood,1999; Glöβler and Milling, 2000; Milling, Thun, 2001; Hatch and Mowery, 2002) |

Kim and Choi, 2009.

*Figure 1. General view of research sector according to the SD theory*

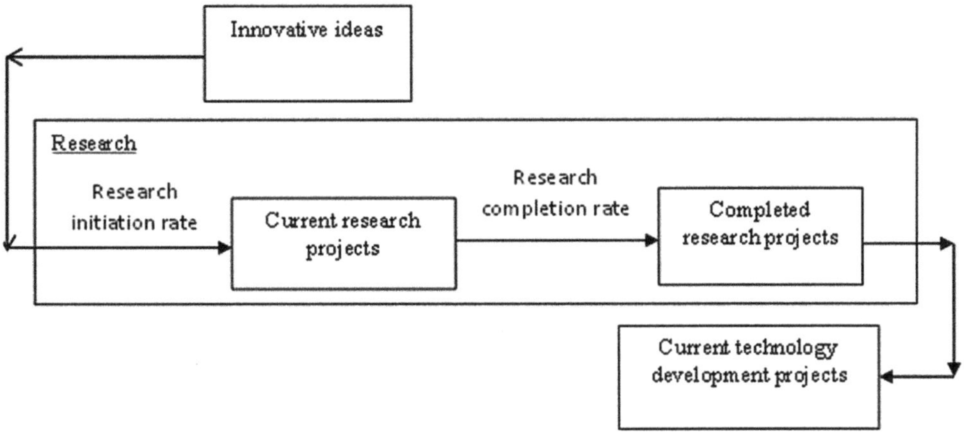

*Figure 2. The effect of resources including human resource, equipment, and budget on time to perform research*

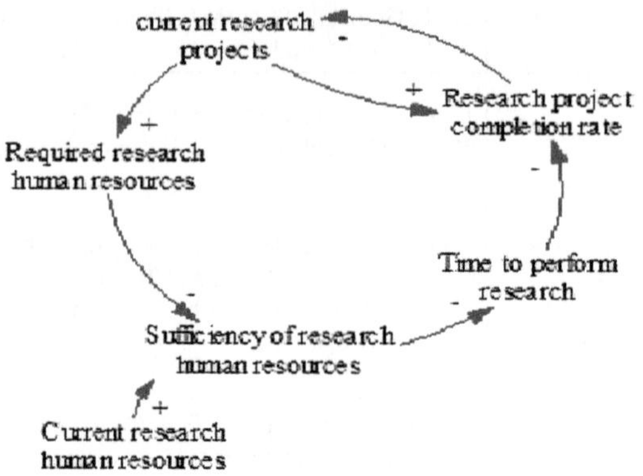


*Figure 3. The impact of experience on time to perform research*

According to Figure 4, an effective factor regarding the initiation rate of research projects is the perception concerning the company's research capacity. Perception of research capacity refers to how managers and researchers recognize the deal of research activities that can be accomplished within a period of one year. In other words, by increasing research human resources, the company's affordance to conduct researches rises. Additionally, the initiation rate of research projects is also influenced by the research project completion rate. Therefore, as the time to perform research increases, initiation rate of research projects and also the number of current research projects will be reduced.

However, the causal-loop diagram needs to be complemented by a quantitative implementation method based on the stock-and-flow diagram. These diagrams can be considered as interfaces between the reality, model and computer simulation. In

*Figure 4. The initiation rate of research projects causal loop*

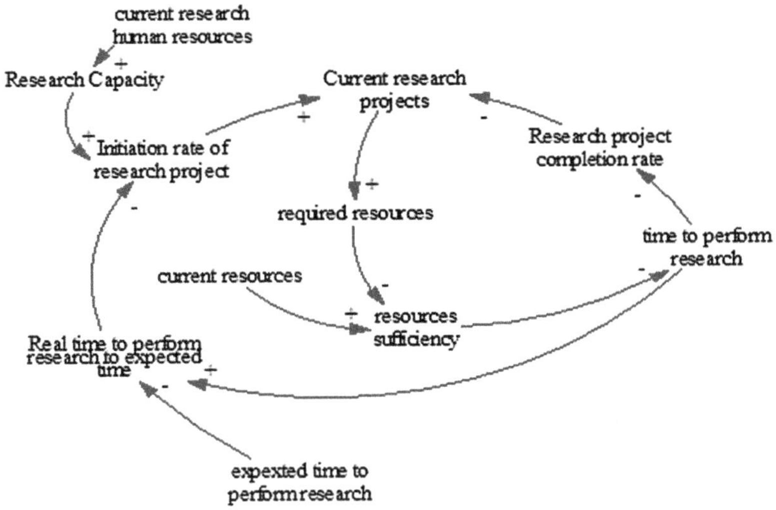


dynamic modeling, variables are used for calculation purposes. Representing the status of a system, such variables as stocks have a measurable value at each time step; flows pointing toward a stock increase this value while flows pointing away from a stock decrease it; auxiliary variables can either contain constants or other parameters. A simplified extract of the dynamic model of the research sector is showed in Figure 5.

Since the purpose of this research is to create technical knowledge, and considering the effect of experience on research activities, experience, as a stock variable, is included in the dynamic model. There are two types of experiences including research experience and market oriented experience with the latter case playing a more important role in creating technical knowledge. The accumulation rate of research experience is determined by the completion rate of research projects. The accumulation rate of market oriented research experience is, however, contingent on the managers' perceived research capacity as well as the completion rate of research projects. Thus, the deal of technical knowledge increases through enhancement of general and, particularly, market related research experiences. Here we only consider stock variable for the research sector:

$$Current\ Research\ Projects = INTEG\left(\begin{matrix} Research\ initiation\ rate \\ -Completion\ rate\ of\ research\ project, 90 \end{matrix}\right) \tag{1}$$

*Figure 5. Dynamic modeling of research sector*

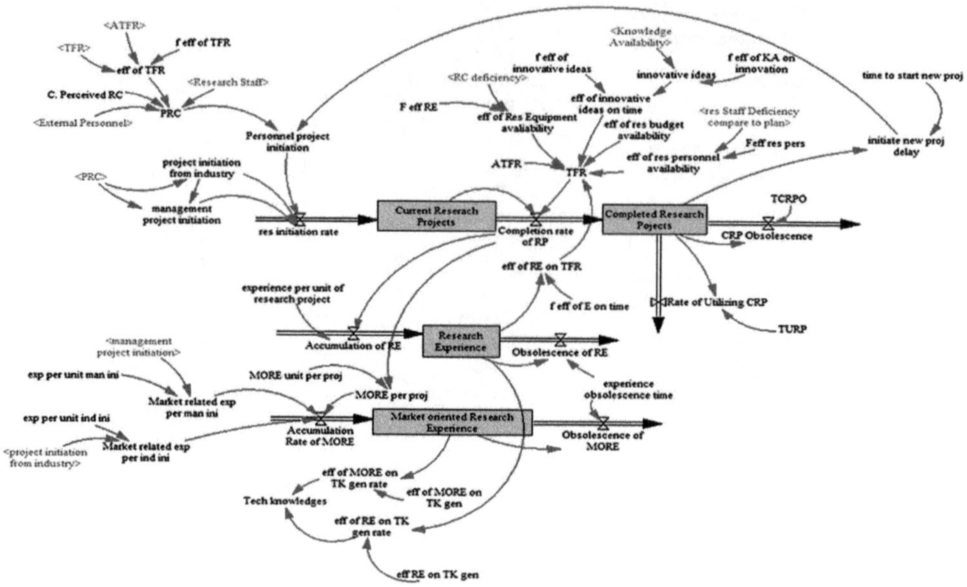


Units: project

$$Completed\ Research\ Projects = INTEG\begin{pmatrix} Completion\ rate\ of\ research\ project \\ -Completed\ research\ project\ obsolescence \\ +Rate\ of\ utilizing\ completed\ reaearch\ project, 500 \end{pmatrix} \tag{2}$$

Units: project

$$Research\ Experience = INTEG\begin{pmatrix} Accumulation\ of\ researchexperience \\ -Obsolescense\ of\ research\ experience, 2000 \end{pmatrix} \tag{3}$$

Units: experience

$$Market\ oriented\ research\ experience$$
$$= INTEG\begin{pmatrix} Accumulation\ rate\ of\ market\ oriented\ research\ experience \\ -Obsolescence\ of\ market\ oriented\ research\ experience, 50 \end{pmatrix} \tag{4}$$

Units: Market related experience

Two important rate variables are as following:

$$\begin{aligned}&\mathit{Completion\ rate\ of\ research\ project}\\&=\left(\mathit{Current\ research\ projects}\right)/\left(\mathit{time\ to\ perform\ research}\right)\end{aligned} \tag{5}$$

Units: project/Month

$$\begin{aligned}&\mathit{Time\ to\ perform\ research(TFR)} = \mathit{Average\ time\ to\ perform\ research}(AFTR)\\&\times\left(\begin{aligned}&\mathit{Effect\ of\ research\ budget\ availability} + \mathit{Effect\ of\ research\ equipment\ availability}\\&+\mathit{Effect\ of\ research\ human\ resource\ availability}\end{aligned}\right)\\&/\left(\mathit{Effect\ of\ innovative\ ideas\ on\ time} + \mathit{Effect\ of\ research\ experience\ on\ time}\right)\end{aligned} \tag{6}$$

Units: Month

ATFR=6

$$\begin{aligned}&\mathit{Research\ initiation\ rate} = 0.4\left(\mathit{management\ project\ initiation}\right)\\&+0.4\left(\mathit{Personnel\ project\ initiation}\right) + 0.2\left(\mathit{Project\ initiation\ from\ industry}\right)\end{aligned} \tag{7}$$

Units: project/Month

## CHALLENGING MODELS WITH SDM IN THE COMPANY

Three methods were used to calibrate and validate the model and to check it in the software (Forrester and Senge, 1979):

- **Test of Model Structure:** Structure verification test,
- **Test of Model Behavior:** Behavior reproduction tests, and
- Behavior prediction tests.

Structure verification means to compare structure of a model directly with the structure of the corresponding real system the model represents. In other words, to pass the structure verification test, the model structure must be consistent with the structure of the real system. Hence it involves comparing model assumptions with those of the managers and decision-makers. In this regard, based on the viewpoints of the general manager, research manager, and the support manager, the model structure is matched to the real life of the company.

Furthermore, behavior reproduction tests examine how well the behavior of the model conforms to observed behavior of the real system. For instance, the symptom generation test examines whether or not a model recreates the symptoms of difficulty that have motivated construction of the model. In this case, the most important problem of the system is the dependency on the research financing. Complying with this problem, the model represents the problem as fluctuations in funding and time to perform research projects which results in a decline in performance. Additionally, model behavior was adapted to the data obtained for the past two years.

Behavior prediction tests examine whether a model generates qualitatively correct patterns of future behavior. The model is seen to behave similar to the forecasting patterns by the managers during the coming three years.

From Figure 6 it is clear that time to perform the research has been fluctuating over time, such that it starts with a rise early each year due to budget unavailability but then it leads to a reduction at the end of the year. The reasons for the unavailability of budget are the time consuming process of fund allocation by the parent company, and also the uncertainty associated with the funding provisions for the coming year. Restrictions on the funding will grow over the time, thus, despite providing such resources as research human resources, research equipment, and increased deal of ideas and experiences, time reduction will not be significant. Moreover, as it is shown in Figure 7, completion rate of research projects has been considerably fluctuating within a limited range likewise research initiation rate with slight fluctuations around a certain value. Meanwhile, both rates exhibit similar volatilities occurring due to the effects of funding volatility. Therefore, it is obvious that the dependency on funding is the major problem causing diverse difficulties.

*Figure 6. Time to perform research*

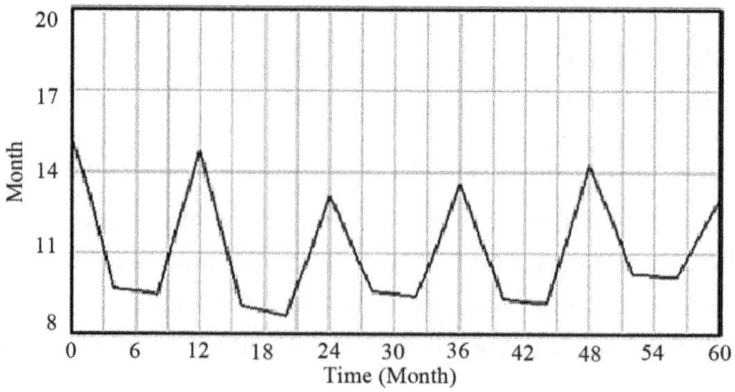

*Figure 7. Research initiation and completion rate*

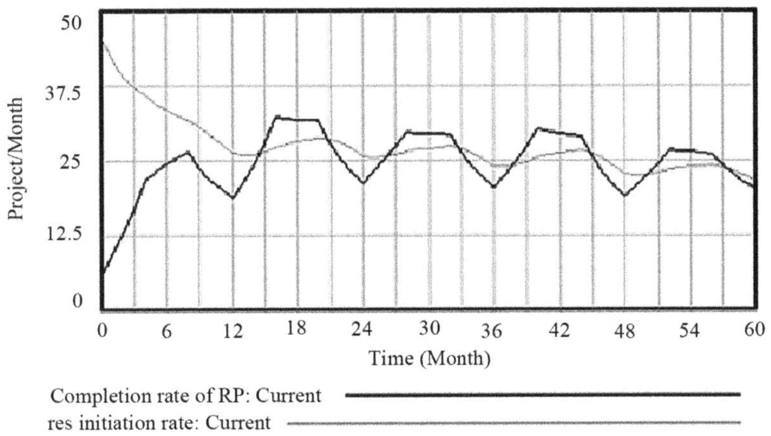


Initially, the company conducted research projects regardless of the market demands. After some time, however, the company redirected to approve proposals in line with the demands raised by market and industry. Moreover, the best way to grow technical knowledge is to concentrate on the market oriented researches. When project initiation (starting point) is concerned, research projects can be divided into three categories: personnel-initiated projects, management-initiated projects, and industry-initiated projects. The latter two deal with researches aligned with the market demands. As is illustrated by Figure 8, during the coming three and a half years, there has been a sharp increase in the number of market oriented researches due to the growth of current and completed research projects as well as the rise witnessed in the number of management- and industry-initiated projects. Since

*Figure 8. Trends about the general research experience and market related experience*

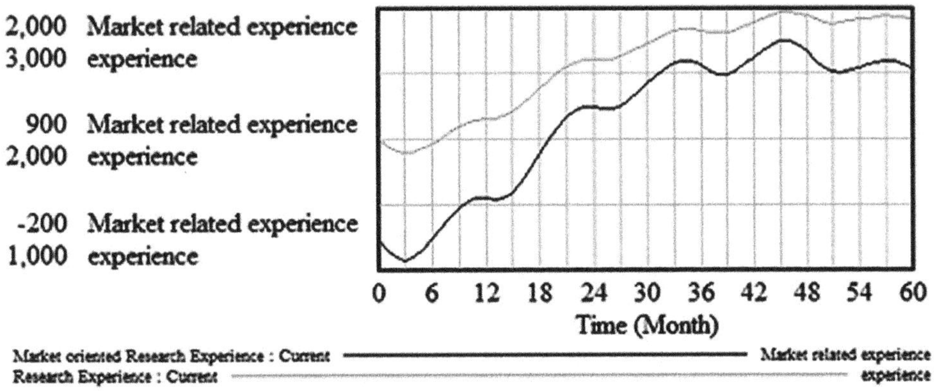

after, it can be seen from the figure that there would be a considerable decline in the market related researches because of the relative stability of current and completed research projects. This process has been repeated for general research experiences. As a result of the budget volatility, both trends have been fluctuating over the years. In addition, after one year, the difference between general research experiences and market related ones is considerably declined as a consequence of a slight increase in the number of management- and industry-initiated projects.

From Figure 9, it is clear that following the first year, the number of both current and completed research projects have been slightly fluctuating around a certain value because of the volatility of the budget.

As is shown in Figure 10, during the coming three and a half years, there will be a significant enhancement in the technical knowledge due to the growth in market oriented and general research experiences. Following this time, it can be seen from the figure that there will be a slight decline in technical knowledge because of considerable decline in market related and research experiences.

## SCENARIO PLANNING WITH SDM IN THE COMPANY

At first, the dependency on the budget should be reduced as it contributes to both initiation and completion rates.

Budget fluctuation causes a lot of volatilities in the research:

*Figure 9. Current and completed research projects*

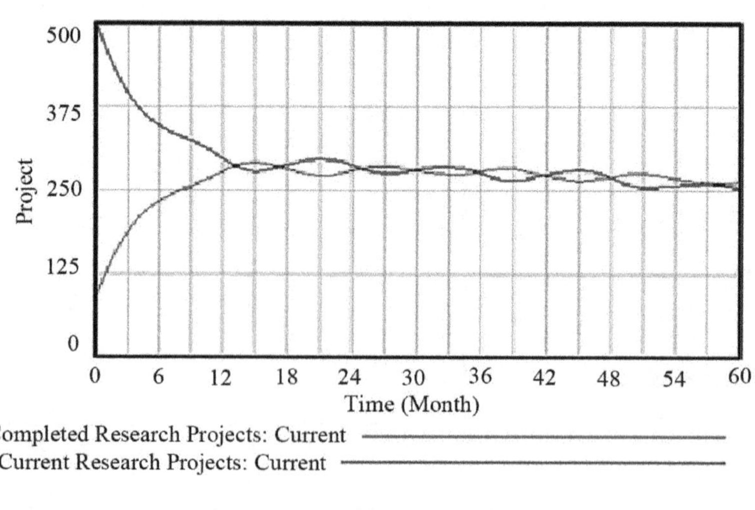

*Figure 10. Technical knowledge*

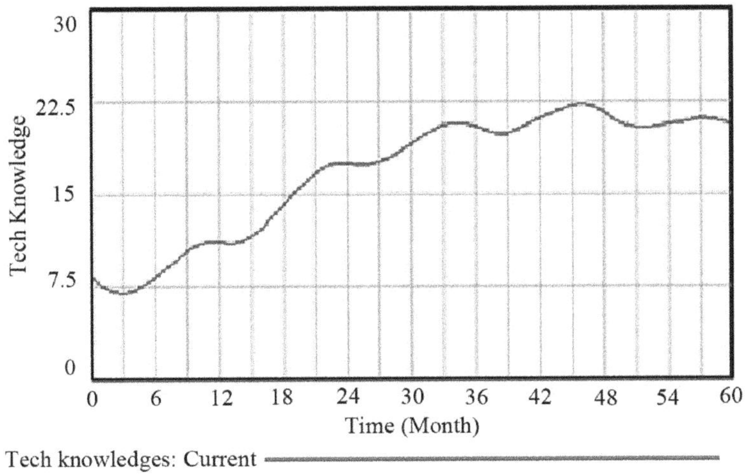


- Time to perform research (TFR),
- Completion rate of research projects,
- Completed research projects,
- Research initiation rate,
- Perceived research capacity,
- Current research projects,
- Outsourcing and external personnel,
- Research equipment deficiency,
- General research experience and market oriented research experience, and
- Technical knowledge.

Consequently, it is proposed, as a policy, to reduce the effect of the funding dependency by 10 percent per year. This policy is suggested to be applied after 2 years.

By virtue of applying this policy, the company can reduce the time to perform research to at most 10 months (instead of 13 months) in three years. That is to say, about 30 percent reduction in the time to perform research (Figure 11). By the same token, research initiation rate will rise by 18 percent while completion rate will grow by 27 percent. Additionally, the trends' fluctuation will be slightly attenuated. Therefore, through reducing the effect of budget availability, the initiation and completion rates will be increased at the same time (Figure 12).

Since the main objective of the company is to enhance the technical knowledge, the company has decided to create forty technical knowledge packages in three years. In other words, company should undertake a scenario under which the technical knowledge is grown by 135% instead of 20% (Figure 13). That is, the number of

*Figure 11. Time to perform research after applying policy*

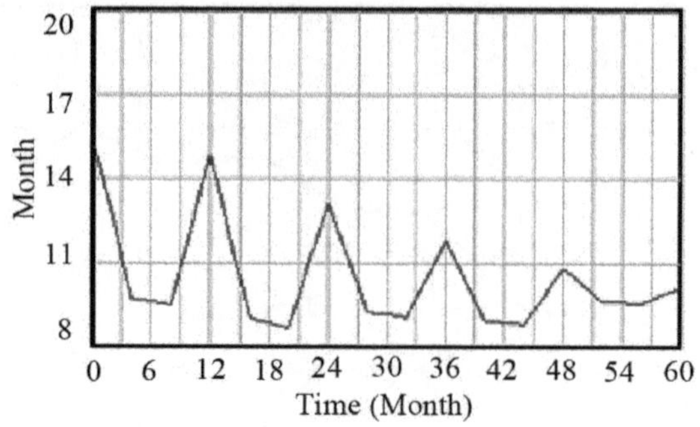


*Figure 12. Research initiation and completion rate after applying policy*

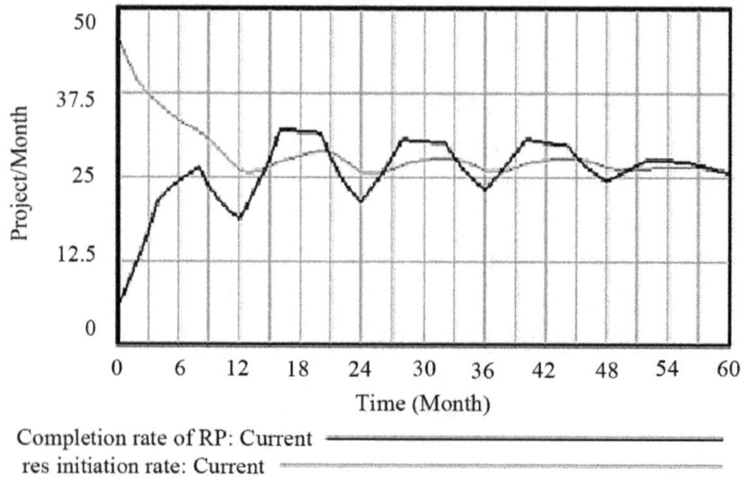


research projects should be raised to 328 projects – about 30% higher than the current number (Figure 14). For this purpose, besides reducing the dependency on the funding at an annual rate of 10 percent, it is proposed to boost research capacity by about 100% to double the number of industry- and management-initiated projects.

Under this scenario, the company needs 266 external personnel, 104 research staffs, and 255 required apparatuses for research activities; this means:

- 34% increase in external personnel,
- 8% increase in research equipment,
- No change in research staff due to the organizational structure.

*Figure 13. Technical knowledge after applying the scenario*

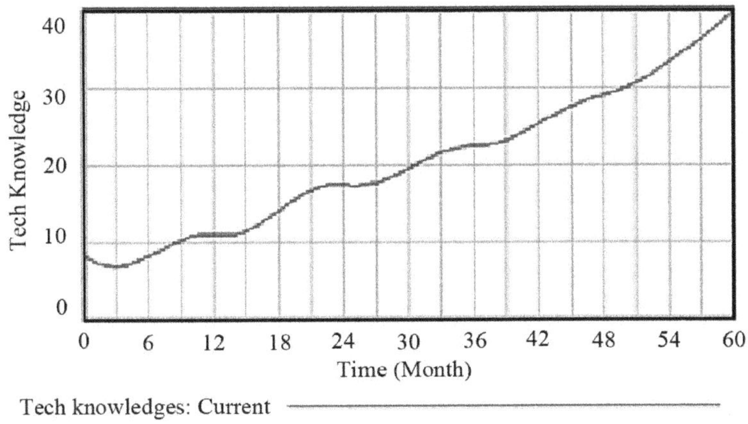


*Figure 14. Current research project after applying the scenario*

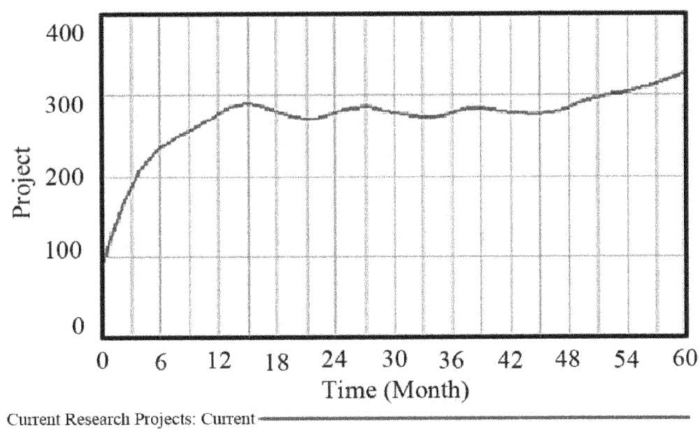


## CONCLUSION

Analyzed in this study was technical knowledge development as the main aim at which Iran's petrochemical industry is oriented. In other words, instead of just importing technologies, a research system is discussed which might be expanded from library research to laboratory investigations and then to bench scale. For this purpose, SD was applied in this study. Although developing technological capabilities and capacities are widely recognized as key drivers of sustainable economic growth, but the research results indicate several obstacles along the path to develop technical knowledge. These include economic barriers, lack of knowledge and expertise, institutional and structural barriers, and lack of understanding about local needs. To overcome these difficulties, a revolutionary movement is required to improve overall infrastructure.

Considering the fund deficiencies particularly evident in the developing countries, it is difficult to make a balance across technical knowledge development and fulfillment of regular activities. Another main factor contributing into technical knowledge development is the firm's access to financial resources. In this case, the most important problem is related to the budget allocation as it influences multitude volatilities and uncertainties within the system. Therefore, the proposed scenario is to eliminate harmful effects of lack of funding or delays in the fund allocation. The firm's budget is composed of the approved budget allocated by the parent company plus the company's revenue. Thus, the company should attempt to grow its revenue. To accomplish this thread, the company should pay attention to the industry's demands and the required technical knowledge.

Being a newcomer in the field of technical knowledge development, Iran has development plans and human resources as its major success factors within research and development activities. The presence of experts in various required fields reduces the dependency of the company on foreign companies for their technologies, so as to increase ROI of the research and development activities. Developing human resource causes a greater deal of competence and experience. As the research results show, time to perform research has been declining sharply over time in view of the fact that expertise and experience have been growing accordingly.

In addition, such entrepreneurial characteristics as risk taking, creativity and innovation are necessary for the researchers. The ability to attract and retain talented technical employees is a critical success factor for companies pursuing technology-based competitive strategies. In this case, experts usually tend to undertake rapidly resulting innovations which can provide significant financial rewards while avoiding to take an easy path and stubbornly follow significant technological breakthroughs. Among other mechanisms, offering interesting conditions, designing reward systems, and nurturing the corporate culture are important mechanisms to attract and maintain these employees. Moreover, to enhance the human resource productivity which contributes to the research initiation rate, it is necessary for the organization to motivate the employees.

In order to enhance researches in terms of reduced time to perform research projects, it is recommended to expand the cooperation with universities and other R&D centers. Furthermore, with the help of communication systems, one can facilitate interactions and co-works across researchers in various industries, universities, research centers, engineering companies, and the company's supporters.

## REFERENCES

Abada, I., Briat, V., & Massol, O. (2013). Construction of a fuel demand function portraying interfuel substitution, a system dynamics approach. *Energy*, *49*, 240–251. doi:10.1016/j.energy.2012.10.063

Barr, P. S., Stimpert, J. L., & Huff, A. S. (1992). Cognitive change, strategic action, and organizational renewal. *Strategic Management Journal*, *13*(S1), 15–36. doi:10.1002/smj.4250131004

Bohn, R. E. (1994). Measuring and managing technological knowledge. *Sloan Management Review*, *36*, 61–73.

Buytendijk, F., Hatch, T., & Micheli, P. (2010). Scenario-based strategy maps. *Business Horizons*, *53*(4), 335–347. doi:10.1016/j.bushor.2010.02.002

Chermack, T. J. (2005). Studying scenario planning: Theory, research suggestions, and hypotheses. *Technological Forecasting and Social Change*, *72*(1), 59–73. doi:10.1016/S0040-1625(03)00137-9

Doyle, J. K., & Ford, D. N. (1998). Mental models concepts for system dynamics research. *System Dynamics Review*, *14*(1), 1, 3–29. doi:10.1002/(SICI)1099-1727(199821)14:1<3::AID-SDR140>3.0.CO;2-K

Feng, Y. (2012). System dynamics modeling for supply chain information sharing. *Physics Procedia*, *25*, 1463–1469. doi:10.1016/j.phpro.2012.03.263

Forrester, J. W. (1971). Counterintuitive behavior of social systems. *Technology Review*, *73*, 52–68.

Forrester, J. W. (1992). Policies, decisions and information sources for modeling. *European Journal of Operational Research*, *59*(1), 42–63. doi:10.1016/0377-2217(92)90006-U

Forrester, J. W., & Senge, P. M. (1979). *Tests for building confidence in system dynamic models*. Retrieved June 13, 2015, from http://clexchange.org/ftp/documents/Roadmaps/RM10/D-2926-7.pdf

Gentner, D. (2002). Psychology of mental models. In N. J. Smelser & P. B. Bates (Eds.), *International Encyclopedia of the Social and Behavioral Sciences* (pp. 9683–9687). Amsterdam: Elsevier Science.

Glick, M. B., Chermack, T. J., Luckel, H., & Gauck, B. Q. (2012). Effects of scenario planning on participant mental models. *European Journal of Training and Development*, *6*(5), 488–50. doi:10.1108/03090591211232066

Goldvarg, E., & Johnson-Laird, P. N. (2001). Naive causality: A mental model theory of causal meaning and reasoning. *Cognitive Science*, *25*(4), 565–610. doi:10.1207/s15516709cog2504_3

Greiner, R., Puig, J., Huchery, C., Collier, N., & Garnett, S. T. (2014). Scenario modeling to support industry strategic planning and decision making. *Environmental Modelling & Software*, *55*, 120–131. doi:10.1016/j.envsoft.2014.01.011

Hoffmann, B., Häfele, S., & Karl, U. (2013). Analysis of performance losses of thermal power plants in Germany - A System Dynamics model approach using data from regional climate modeling. *Energy*, *49*, 193–203. doi:10.1016/j.energy.2012.10.034

Khodabakhshian, A., Khosravi, M., & Mashayekhi, A. N. (2013). R&D project portfolio in research and development, part of a whole in R&D strategy. *System Dynamics Conference*. Retrieved August 30, 2017, from http://www.systemdynamics.org/conferences/2013/proceed/papers/P1299.pdf

Kim, S. W., & Choi, K. (2009). *A Dynamic Analysis of Technological Innovation Using System Dynamics*. POMS 20th Annual Conference, Orlando, FL. Retrieved August 30, 2017, from https://www.pomsmeetings.org/ConfPapers/011/011-0622.pdf

Kolkman, M.J., Kok, M., van der Veen, A. (2005). Mental model mapping as a new tool to analyse the use of information in decision-making in integrated water management. *Physics and Chemistry of the Earth*, *30*(4-5), 317–332.

Kortea, F. R., & Chermack, T. J. (2007). Changing organizational culture with scenario planning. *Futures*, *39*(6), 645–656. doi:10.1016/j.futures.2006.11.001

Kusumaningdyah, W., Eunike, A., & Yuniarti, R. (2013). Modeling tradeoff in ship breaking industry considering sustainability aspects: A system dynamics approach. *Procedia Environmental Sciences*, *17*, 785–794. doi:10.1016/j.proenv.2013.02.096

Lall, S. (1993). Promoting technology development: The role of technology transfer and indigenous effort. *Third World Quarterly*, *14*(1), 95–109. doi:10.1080/01436599308420315

Lane, D. C. (2000). Should system dynamics be described as a hard or deterministic systems approach. *Systems Research and Behavioral Science System*, *17*(1), 3–22. doi:10.1002/(SICI)1099-1743(200001/02)17:1<3::AID-SRES344>3.0.CO;2-7

Lee, C.-F., & Chung, C.-F. (2012). An inventory model for deteriorating items in a supply chain with system dynamics analysis. *Procedia: Social and Behavioral Sciences*, *40*, 41–51. doi:10.1016/j.sbspro.2012.03.159

Leonard-Barton, D. (1995). *Wellsprings of knowledge*. Boston: Harvard Business School Press.

Li, A., & Maani, K. (2011). *Dynamic Decision-Making, Learning and Mental Models*. Retrieved August 10, 2014, from http://www.systemdynamics.org/conferences/2011/proceed/papers/P1189.pdf

Liuguo, S., Shijing, Z., & Jianbai, H. (2012). Pricing simulation platform based on system dynamics. *Systems Engineering Procedia*, *5*, 445–453. doi:10.1016/j.sepro.2012.04.067

Marcus, G. P. (1992). Present and emergent identities: requirements for ethnographies of late twentieth-century modernity worldwide. In S. Lash & J. Friedman (Eds.), *Modernity and identity* (pp. 309–330). Oxford, UK: Blackwell.

Milling, P. M., & Maier, F. H. (2009). Research and development, technological innovations and diffusion. *System Dynamics*. Retrieved August 30, 2017, from http://www.eolss.net/sample-chapters/c15/E6-63-01-06.pdf

Monga, P. (2001). *A System dynamics model of the development of new technologies for ship systems*. Master of Science in Industrial and Systems Engineering. Retrieved August 30, 2017, from https://theses.lib.vt.edu/theses/available/etd-10022001-014531/unrestricted/Pavinder_Monga_Masters_Thesis.pdf

Morecroft, J. D. W. (1988). System dynamics and micro worlds for policymakers. *European Journal of Operational Research*, *35*(3), 301–320. doi:10.1016/0377-2217(88)90221-4

Park, S., Kim, B.-J., & Jung, S.-Y. (2014). Simulation methods of a system dynamics model for efficient operations and planning of capacity expansion of activated-sludge wastewater treatment plants. *Procedia Engineering*, *70*, 1289–1295. doi:10.1016/j.proeng.2014.02.142

Rasjidin, R., Kumar, A., Alam, F., & Abosuliman, S. (2012). A system dynamics conceptual model on retail electricity supply and demand system to minimize retailer's cost in eastern Australia. *Procedia Engineering*, *49*, 330–337. doi:10.1016/j.proeng.2012.10.145

Rodriguez-Ulloa, R., & Paucar-Caceres, A. (2004). *Soft system dynamics methodology (SSDM): A combination of soft systems methodology (SSM) and system dynamics (SD)*. Retrieved August 6, 2015, from http://www.systemdynamics.org/conferences/2004/SDS_2004/PAPERS/163PAUCA.pdf

Saliba, G. (2009). Windows for the mind: The use of scenario planning for enhancing decision-making and managing uncertainty. *Journal of Futures Studies*, *13*(3), 123–129.

Senge, P. M., & Sterman, J. D. (1992). Systems thinking and organizational learning: Acting locally and thinking globally in the organization of the future. *European Journal of Operational Research*, *59*(1), 137–150. doi:10.1016/0377-2217(92)90011-W

Tushman, M. L., & Romanelli, E. (1985). Organizational evolution: A metamorphosis model of convergence and reorientation. *Research in Organizational Behavior*, *7*, 171–222.

Wang, W. (1997). Technology transfer to and absorption in a developing world—case studies of Ericsson in Malaysia and China. *Sixth International Conference on Management of Technology*, 25–28.

Wilner, N., & Birnberg, J. (1986). Methodological problems in functional fixation research: Criticism and suggestions. *Accounting, Organizations and Society*, *11*(1), 71–80. doi:10.1016/0361-3682(86)90019-X

Zahradnícková, L., & Vacík, E. (2014). Scenarios as a Strong Support for Strategic Planning. *Procedia Engineering*, *69*, 665–669. doi:10.1016/j.proeng.2014.03.040

Zaim, S., Bayyurt, N., Tarim, M., Zaim, H., & Guc, Y. (2013). System dynamics modeling of a knowledge management process: A case study in Turkish Airlines. *Procedia: Social and Behavioral Sciences*, *99*, 545–552. doi:10.1016/j.sbspro.2013.10.524

Zhang, H., Calvo-Amodio, J. R., & Haapala, K. (2013). A conceptual model for assisting sustainable manufacturing through system dynamics. *Journal of Manufacturing Systems*, *32*(4), 543–549. doi:10.1016/j.jmsy.2013.05.007

Zollo, M., & Winter, S. G. (2002). Deliberate learning and the evolution of dynamic capabilities'. *Organization Science*, *13*(3), 339–351. doi:10.1287/orsc.13.3.339.2780

# Chapter 4
# Using Modeling and Simulation Techniques to Improve the Accountability and the Performance of Human Resource Management

**Florea Nicoleta Valentina**
*Valahia University of Targoviste, Romania*

**Manea Marinela Daniela**
*Valahia University of Targoviste, Romania*

## ABSTRACT

*The analysis of human resources function and its contribution to obtain performance dates back to the 1920s. Now, the HR is an equal partner on the board of the companies, having a strategic role in obtaining performance, thus we try to show that compensating appropriately, the human capital it will be motivated to obtain performance. This paper examines the two different visions of different managers in which the human capital is perceived as a major cost for organization and the others which perceive it as an investment on long run. In this article, we analyse the impact could have the direct costs of human capital on individual and organizational performance using samples of some variables from European level, data between 2005-2016. Data used for the different years were analysed using simulation methods. Findings of this study show consistency with the theory in the filed, bringing a value in motivation and accountability of human capital and performance obtained through human capital.*

DOI: 10.4018/978-1-5225-3226-2.ch004

## INTRODUCTION

Human capital became crucial for any organization in obtaining performance. Due to its unique and valuable skills, knowledge, and experience, human capital is a key and strategic resource for achieving competitive advantage, so it is for many organizations an investment and not a cost. Thus, human capital analysis, controlling and accountability gain great importance in performance organizations, being the largest cost from the operating costs. And because into the external environment there is a continuous change, every step it is met the globalization and the liberalization of the markets, the organizations must take into account the possible risks, the high costs, the uniqueness of human capital, the nature and the structure of work, the needs of employees and of the customers, the labor conditions and to use some special techniques necessary to predict the appearance of these risks: simulation and modeling. These tools will help organizations to know the next future using past data about the organization, necessary to forecast future problems linked, in our case, by the human capital.

Human capital is unique, having unique skills, attitudes, behaviors, knowledge, intelligence and capacity to use tools, methods and instruments necessary to improve any process, and obtain individual and implicitly organizational performance. To maintain on long term the talented human capital, organizations must implement some practices in order to attract this capital, to retain it and to use it by developing it continuously. To do these, the smart organizations may use best HR practices as financial motivation or compensation, because this is the most visible process outside organization, which help it retain talented staff on long run. Compensation is consisting in direct and indirect costs for human capital.

This paper wants to offer an answer to an important issue: "Does direct costs of human capital influence the capacity of individuals and organization to obtain motivation, work satisfaction and performance?" In order to answer this issue, we collected data from European level, from 28 countries and denoted as EU (28 countries) between 2005-2016 regarding some variables which are presented into our research methodology.

Our study is offering in the first part the literature review on human capital and its role in obtaining performance on long term. The second part is analyzing the human capital from two points of view: that old point of view where the capital was perceived as being a cost and the second one which is as being an investment and a major factor in adding value and obtaining competitive advantage. The third part presents the influence could have human capital direct costs on the variable presented above, using simulation and we present the research methodology. The results, discussions, proposals and measure to improve the study are provided in the fourth part. Finally, the fifth part concludes the study.

Our objective is to find a relationship between the money spend on human capital (compensation) and the results they obtained. We will use data regarding the human capital expenses to show the importance of measuring them, finding correlations between different sets of HR data using simulation techniques, and also determining the appropriate amount level for human resources compensation, in order to be motivated and satisfied and retained on long run.

## BACKGROUND

Any organization depends on three important factors: human capital, processes and technologies, the first factor being the success factor (Reece et al., 2011, p.5). Human capital is an important and valuable resource for any organization in order to obtain performance. They have to use performance processes in order to attract, to retain on long term and to satisfy talented human capital. Doing these, the organization will ensure its success on the globalized markets. Sims (2002, p.5) says that this resource lead organizations to reach their objectives and to obtain performance. In the vision of Reed (2001, p.1) human factor is recognized today as a central factor and the most important resource necessary for the business and in the opinion of Faulkner (2002, p.136) it is a key resource, vital for organizations in obtaining competitive success and the most important that an organization may have. More and more the competitive advantage resids in the organization' employees. DeCenzo et al. (1988, p.6) says that an organization is nothing without its people, eventually being some expensive equipment. Thus, from this points of view human capital is representing one of the most important investment of an organization, which results are becoming much evident in time (Manolescu, 2006, p.35), the fundamental element of an organization succeeding (Lakhdar S. şi col.,2001, p.51), a valuable asset of an organization, who contribute to the organization performance (Armstrong& Baron, 2003, p.107) and one of the hidden values, who create value and future power (Brewster& Larsen, 2007, p.1). When an employee is satisfied by the conditions the organization offers, the more that employee identifies with the organization, the more engaged they are likely to be and the less likely to look for other employment (Moynihan and Pandey, 2007). Thus, organizations must fight for its talented employees, otherwise they will lose the young employees who are more socialized to the new employment contract and more likely to change positions (Coleman Selden, 2008).

The Government Performance Project (2007) found that voluntary turnover correlated significantly with both average age and years of service. Governments with an older workforce and with greater average tenure had significantly less voluntary turnover. Human capital practices, such as compensations, benefits, quality of supervision, training, development, rewards, recognition, performance feedback, and job design, also influence voluntary turnover. Particular job design attributes, such as responsibility, task variety, role clarity, and autonomy, are also commonly associated with greater job satisfaction and less turnover.

However, Bohlander & Snell (2007, p.4) argue that human capital is more then ever important into the organizations, and Millmore et al. (2007, p.15) sustain that is representing a key resource and influencing every component of the strategic management of an organization. In order to achieve its objectives, organizations must match its employees with the organization and work together in order to achieve its objectives (Palmer, 2012, p.135). Akay (2006, p.5) argue that human resource is the link between the four pillars represented by the company, the management, the communication and the production offering the strength to stand up and sustain the entire business.

As a final and important conclusion is that the organizations, if they will not invest in its human capital, probably they will not be able to develop and retain the key-resources in the next years (Holbeche & Mayo, 2012, p.30). Competitive advantage will be obtained by those organizations who are able to attract, select, retain and develop the taleted people (Bohlander & Snell, 2007, p.4), and efficient retaining may be done using financial or non-financial motivation. Financial motivation supposes offering salaries, wages, bonuses, or incentives.

The concept of human capital advantage as formulated by Boxall & Purcell (2007) is based on the belief that sustainable competitive advantage is achieved when the firm has a HR pool that cannot be imitated or substituted by its rivals (Barney, 1991). Unique talents among employees, including superior performance, productivity, flexibility, innovation and the ability to deliver high levels of personal customer service are ways in which people provide a critical ingredient in developing an organization's competitive position. People also provide the key to managing the pivotal interdependencies across functional activities and the important external relationships (Baron& Armstrong, 2007, p.26).

Human capital to be competitive, it must be valuable, rare, hard to imitate by the competition and non-substitutable. It can be argued that one of the clear benefits arising from competitive advantage based on the effective management of human capital is that such an advantage is hard to imitate (Baron& Armstrong, 2007, p.26). An organization's HR strategies, policies and practices are a unique blend of processes, procedures, personalities, styles, capabilities and organizational culture. One of the keys to competitive advantage is the ability to differentiate what the business supplies to its customers from what is supplied by its competitors.

Talentship decision frameworks are important within HR as well as outside HR. The need for a talentship mind-set is also vital for HR functional leaders. In our experience there are almost always some HR professionals in any organization who effectively understand, teach, and enhance decisions based on how talent connects to strategic success (Storey, 2009, p.31). These HR professionals typically admit that they learned to provide this kind of support in their own way, with little systematic instruction or development.

## ISSUES, CONTROVERSIES, CONSEQUENCES

About the Role of Modeling and Simulation Techniques in HRM Accountability

*The best way to predict the future is to create it. – Abraham Lincoln*

To measure performance of human capital, the organizations may use some techniques necessary to monitor, analyse and value this strategic asset. Among the analysis made: statistical process control, brainstorming, problem analysis, cause-and-effect diagram, force field analysis, mind mapping, affinity diagrams, diagnostic instruments, focus groups, probing interviews, job satisfaction surveys, engagement surveys, exit interviews and surveys, nominal group techniques (Phillips J.J. et al., p.91), and also modeling and simulation techniques. These challenges may be analyzed using different tools and techniques, which may easy the acquisition of new knowledge, such as: paper images, videous images or software applications (Sa Silva et al, p.63 from EOMAS 2011). The organizations do not allow intuitive actions to develop, that is why the simulation using computer became important more than ever (McHaney R., 2009, p.9).

Multiple simulation and modeling techniques were developed in order to analyse complex systems and solve problems related to complex issues and are many, but from all we reminding just a few:

- Systems dynamics (SD) modeling,
- Discrete event simulation (DES),
- Agent-based modeling (ABM),
- Monte Carlo simulation- OLS method.

To understand why was chosen the OLS method in our research, we offer some insights and points of views of specialists in the field.

- **Systems Dynamics:** SD is based on differential equations and tries to capture systems with the so-called stock-and-flow-notation. Both SD and ABM are regularly used to explain socio-economical phenomena but differ in the way they approach their explanandum. SD looks for a reference mode of the central variable, whereas in ABM and agent is modeled and the emergent behaviour arising out of the interaction of a population of those agents is being observed (Squazzoni F., 2009, p.145).

- **Discrete Event Simulation:** The increasing popularity of DES models (in applications and in research) suggests that it exists at the right level of definition (Wainer G., 2016, p.xviii). DES can play a vital role in resolving issues using the analytically based optimal policy and execute it using other management policies (Fishman G.S., 2013, p.24, 26). DES on a digital computer offers many conveniences that make it an attractive tool for analysis:
    - It compresses time allowing to run through several competing operational designs in a very small fraction of the time,
    - It expands the time by arranging for statistics of interest to be produced over small intervals of simulated time, allowing an investigator to change later what can not see in real time,
    - It offers the ability to identify and control sources of variation,
    - Maintain a list of events scheduled to be executed as time evolves,
    - Maintain other dynamical data structures,
    - Generate pseudorandom numbers,
    - generate samples form a variety of distributions,
    - compute summary statistics including time averages, variances, and histograms,
    - generate reports and detect errors.
- **Agent-Based Modeling:** ABM allows some systems to be represented in a more rational and comprehensible way than would be the case with other simulation techniques. In DES we focus on how entities pass through a network of queues and activities, in SD we focus on stocks (entities), their flows and their interdependence, and ABM we focus on the agents and their relationships with each other and their environment. ABM typically has four aspects (Taylor S., 2014):
    - A set of autonomous agents with a set of attributes and specified behaviours, and define how an agent behaves in response to changes and towards a set of goals,
    - A set of agent relationships, each defining how each agent interacts with other agents and its environment,
    - The agent's environment where is existing,
    - A system is therefore composed of the set of agents, the environment and their relationships. A system has a clearly defined boundary with well-defined inputs and outputs.

ABM is individual based but the focus is on modeling the entities and interactions between them. DES is useful for problems that consist of queuing simulations or complex network of queues, in which the process can be well defined and their emphasis is on representing uncertainty through stochastic distributions (Rodriguez J., et al., 2012, p.78).

- **Monte Carlo Simulation:** A practical method for checking properties of the estimators (OLS estimator- among all linear estimators OLS has the smallest variance and in this sense is the best linear unbiased estimators (BLUE). The specialists have completed control of the way data is generated (they set the values of the unknown parameters from the outset) (Seddighi H. et al., 2000, p.34). OLS is a discrete event simulation framework (Prasad A. et al., 2011, p.110) and use the probability calculations (Florea N.V., Mihai D.C., 2017, p.89-90).

If all assumptions of the model are consistent with the data, then OLS estimators (p.33):

- Are unbiased (the expected value of the sampling distribution is equal to the unknown parameter),
- Have minimum variance (this property is conditional on the validity of the assumption on non-autocorrelation and homoscedasticity).

Given the properties mentioned above, OLS estimators are said to be efficient, and the efficiency depends on the validity of the model's assumptions (if any of the assumptions fail, then the OLS method is no longer efficient and another estimation methods must be found) (Seddighi H. et al., 2000, p.33). The aim of OLS method is to generate efficient estimators of the unknown parameters so that each parameter can be estimated given a set of observations on the dependent and independent variables. To estimate parameters of a simple model is used linear regression model using available data from random sample. OLS is by far the most used method developed by Gauss. The method is chosen due its reduced costs, to the availability of the data, its aim being to minimize the sum of squared deviations with respect to the parameters (Lawal B., 2014, p.219). The aim of regression analysis is (Seddighi H., 2013, p.44): to estimate econometric models, to provide empirical content, to abstract economic theories and models, to find estimators capable of estimating the parameters of the regression models, to determine the estimators, using a mathematical formula, to generate point estimates for that unknown parameter on the basis of a set of a set of observations on the dependent and independent variables), to analyse the relationship strength between the analysed variables (Florea N.V., 2015, p.375; Florea N.V., Mihai D.C., 2015, p.233; Tanasescu D.A. et al., 2014, p.34; Duica A. et al., 2016, p.101) using different programs (Eviews, R or Excel) (Florea N.V., 2017, p.141).

All too often, these developments have taken place indistinct fields, developing into separate "schools" each ascribing dynamic complexity, such as for SD- feedback and accumulation effects or for ABM- heterogenous actor-specific (inter) actions (Janssen M. et al., 2015, p.76).

To understand better the different analysed methods we offer some comparison between them using some characteristics (Table 1).

And because we are surrounded by continuous changing, the organizations want to know their future in advance of some processes from time and also their performance. To realize this thing, the organizations may use simulation and make decisions who not imply costs, based on using data and information from the past

*Table 1. Comparison of SD, ABM, DES, and OLS*

| Characteristics | System Dynamics | Agent-Based Modeling | Discrete Event Simulation | Ordinary Least Squares |
|---|---|---|---|---|
| Basic building block | Feedback loop | Agent | Modular and hierarchical formalism | Relations |
| Unit of analysis | Structure | Rules | Input and output parameters | Phenomenon |
| Level of modeling | Macro | Micro | Macro | Micro and macro |
| Perspective | Top-down | Bottom-down | Top-down | Test prediction |
| Adaptation | Change of dominant structure | Change of structure | Change of framework | Change of results |
| Handling of time | Continuous | Discrete | Discrete events systems | Systematic and stochastic |
| Mathematical formulation | Integral equations | Logic | Logical-analytical | Regression function |
| Origin of dynamics | Levels | Events | Non-centric systems | Relation |
| Structure | Static | Dynamic | Dynamic | Randomly |
| Behavior | Feedback is central | Accumulations are central | Detail is central not entities | Samples of data is central |
| Feedback | Flat | Inter-level feedback | Domain specific languages (queues) | Future patterns |
| Emergence | Mono-level structure | Multilevel structure | All system | Multiple random samples |

Source: Adaptation after Carsey T.M., Harden J.J., 2013, p.6-8; Mittal S. et al., 2017, p.94; Fishman G.S., 2013, p.23-24; Taylor S., 2014.

(OLS method). To achieve these objectives, the organization must develop and use some analytical tools which may be used to model, analyze, simulate and project a business process (Laguna M., Marklund J., 2011, p.98).

Modeling and simulation are necessary when direct experiencing is not possible on the real system. A simulation process may increase the quality of performance, the rational consumption of limited resources, may reduce time and costs, which are important aspects of a present global economy (Banks et al, 1996; Law & Kelton, 1991, p.544). Simulation is used into the following conditions: when experiencing is not possible or when the analytical solution is unknown or uncertain (Davis& Pecar, 2013, p.551), but is used in different contexts, demonstrating flexibility in solving the issue and obtaining viable solutions (Sa Silva P. şi col., p.63 din EOMAS 2011). Now, using simulation is a growing and continuously used tool for different fields (Moore, J.H., 2001).

Certains models can be simulated to determine the costs or the life-cycles, the future results of some strategies or future trends, in order to discover errors or gaps of performance (Bridgeland D.M., Zahavi R., 2008, p.291). Simulation is a powerful tool used to manage the problems and to project the entire organizational system (Sabuncuoglu I. et al., 2015).

Using simulation it may be developed models, who may help specialists to make future estimations about competition, their development plans or their profitability (Powell G.R., 2008). Many organizations understood that simulation became an important tool used to solve different situations and understand them, by collecting data and find solutions in safe conditions (Sokolowski & Banks, 2011, p.3). Simulation is a paradigm sued to analyse complex systems by using simplified representations. This tool may conduct to improve cost-benefit analysis, experiencing, develop new skills and obtain performance (Altiok T., Melamed B., 2010, p.1). And because simulation is a method which may study the performance of a real system (Anderson, Beveridge& Scott, 2000), we propose to implement such a technique to improve organization performance through the prism of human capital efficiency and the monitoring activity of human capital expenses which add value on long run.

## Human Capital: A Cost or an Investment?

An organization must assure its existence on the market in order to become performant and it must know where it is and where it has to go by achieving its objectives. To do that, it has to implement a continuous monitoring activity. To monitor the human capital performance the organization must analyse in the first place the compensation of its employees, because this is the largest costs and from this point of view, it must know its evolution along time.

The human capital expenses have the greatest importance in the structure of expenses, that is why is necessary to know continuously their limit and the established budget.

Developing an effective strategy to manage an organization's human assets requires considering employees as investments (Mello, 2014, p.15). Such an approach helps to ensure that HR practices and principles are clearly in sync with the organization's overall strategy, forces the organization to invest in its best opportunities, and ensures that performance standards are met. As an example, employee stock-ownership programs attempt to strategically invest in the organization and its people by making employees owners of the company. Instead of having a conflict as to how profits should be allocated (bonuses to employees or reinvested in the business), both can be achieved simultaneously. In turn, this has the goal of gaining more commitment from employees and encouraging them to adopt a long-term focus toward the organization; this is often a shortcoming or deficiency of American organizations that are concerned with short-run indicators of performance. Employees who now intend to stay with organization longer, given their vested ownership right, provide organizations with an incentive to incur the short-term costs involved with investing in human assets for the long-term financial gains that can result from such investments.

An investment perspective of HR is often not adopted because it involves making a longer-term commitment to employees. Because employees can "walk" and because American organizations are so infused with short-term measures of performance, investments in human assets, which tend to be longer-term investments, are often ignored. Organizations performing well financially may feel no need to change their investment strategies. Those not doing well usually need a quick fix to turn things around and therefore ignore longer-term investments in people. However, while investments in HR are longer term, once an organization gains a competitive advantage through its employees, the outcomes associated with the strategy are likely to be enduring and difficult to duplicate by competitors as such programs and values become more firmly entrenched in the organization's culture.

In the opinion of Mayo (2001) the essential difference between HCM and HRM is that the former treats people as assets while the latter treats them as costs. Kearns (2005a) believes that in HCM' people are value adders, not overheads' while in HRM' people are [treated as] a significant cost and should be managed accordingly' (Baron & Armstrong, 2007, p.24). The claim that in HRM employees are treated as costs is not supported by the descriptions of the concept of HRM produced by US writers such as Beer at all (1984). In one of the seminal texts on HRM they emphasized the need for' a longer - term perspective in managing people and consideration of people as potential assets rather than merely a variable cost'. Fombrun et all (1984), in the other seminal text, quite explicitly presented workers as a key resource that managers use to achieve competitive advantage for their companies. According to the

Kearns (2005a), in HRM' the HR team it seen as a support service to the line' - HR is based around the function and the HR team performs' a distinct and separate role from other functions'. Conversely, 'HCM is clearly seen and respected as an equal business partner at senior levels' and is 'holistic, organization-wide and systems-based' as well as being strategic and concerned with adding value.

## Compensation of Human Capital: The Largest Expense for Organizations, but the Primary Active in Obtaining Performance

Also compensation and benefit expenses are often the largest individual line item expense in any organizational setting (Bashker, 2014, p.8). Relevant data shows that total compensation expenses in organizations fall within 20% to 60% of gross revenue. In the service sector, this percentage is in the 50% to 60% range. If one considers salaries as a percentage of operating expense, the range can be from 15% to 50%. Data from the Bureau of Labor Statistics in 2008 suggests that in the healthcare industry the salaries to operating expenses ratio was as high as 52%. In for-profit service organizations, the ratio was 50%. In durable goods manufacturing, construction/mining, and oil/gas, the ratio was 22%. And in the retail sector, the salaries to operating expenses ratio was as low as 18%.

Compensation and benefits is the largest expense item for any organization with regard to human capital. Therefore, there is a need to clearly understand and articulate the links between compensation and benefits and accounting (finance). It also suggests a need for a closer alignment of accounting (finance) with the activities of compensation and benefits. Note, as well, that many financial problems can be explained by compensation systems or by the specifics of the tax code. When one cannot explain a firm's behavior with economic logic, the real answer may often lie in compensation systems. We will explore these connections in more detail throughout the book.

This chapter's objective is to determine the relation between the accounting (compensation and benefits) and the organization performance. HR-related accounting and finance implications are usually captured in accounting and finance texts in an unconnected manner. In contrast, this book brings into focus in one single publication all of these compensation and benefit and accounting (finance) topics, discussing the major compensation and benefit sub-functions one by one. Within each subtopic, you learn the relevant accounting and finance implications.

Throughout this chapter, the compensation and benefit topics that have major accounting implications are discussed, and referring to human capital. When talking about compensation and benefits, you must consider that a total compensation program consists of various elements. Normally, a total compensation structure includes the following elements: Base, Cash incentives or bonuses, Equity compensation, Cash-based long-term incentives, Executive compensation, Sales compensation, Expatriate compensation, Risk benefits, Retirement benefits, Perquisites, Other Benefits

The words costs and expenses are used interchangeably in accounting. But a cost incurred can be an asset or expense depending on the timing of accounting transactions and the concept of periodicity. Especially in transactions like the acquisition of a physical asset, the cost classification can become an important decision. When a physical asset is acquired, many costs might be involved (for example, purchase price, freight costs, and installation costs). So, the accountant has to decide which cost to include as an asset and which costs to expense immediately. Those costs that are expensed immediately can be called revenue expenditures. And costs that are not expensed immediately but are included in asset accounts are referred to as capital expenditures. Some firms call these expenses operating expenses (OPEX) and capital expenses (CAPEX).

An expense is, in actuality, a cost used up while producing the sales revenue for the business. In other words, expenses are those monetary outlays that flow through to the income statement. In contrast, costs that have not been used up remain a cost and are reported on the balance sheet as an asset. Compensation has long been considered one of the most important organizational rewards because it allows employees to obtain other rewards (Lawler, 1971). Compensation management entails a mechanism that connects employee efforts, performance and outcomes with organizational goals and objectives. Since a large number of employees work in an organization, therefore, managing compensation in the organizational context involves many considerations for designing and implementing rewards for employee's work and performance, which needs to be developed for maintaining motivation and productivity of employees on one hand organizational effectiveness on other hand (Tapamoy, 2009, p.107). These considerations in compensation management require managerial decision-making and appropriate planning. According to Hackett and McDermott (1999), considerations in compensation management are presented in Table 2.

Thus, organizations need to decide on the above issues for ensuring employee efficiency and organizational cost control, which can be broadly classified as under:

- Level of compensation is adequate to attract and retain competent employees necessary for creating competitive advantage for the organization,
- Structure of compensation to facilitate the attainment of an organization's strategic goals and objectives,
- System of compensation is perceived by employees as fair and equitable and complies whit government regulations,

*Table 2. Compensation considerations*

| Issues | Things to Consider |
|---|---|
| Business strategy | Is the compensation approach in synchrony with the organization's strategic plan? |
| Corporate culture | If the organization is expecting to pay for performance, do managers and employees have the willingness and ability to share the information necessary to improve their performance? |
| Compensation philosophy | Does the organization pay below, at or above market? |
| Pay for performance | Is management willing to provide above market pay for superior performance? |
| Base pay | Does the organization prefer to pay at market or use merit approach? |
| Compensation structure | How will the organization structure its compensation program - no ranges, traditional ranges or broad banding? |
| Variable pay | Which is the most appropriate for the organization, individual, group, team or organization wide incentives? |
| Performance management | Is the organization able to develop good job, team or group performance measures to support any new compensation approach? |
| Promotion/Turnover | Is the organization willing to pay a premium to retain highly skilled human resources? |
| Individual compensation programs | Does the nature of the work lend itself to skill or merit, competency or knowledge or marked-based pay systems? |
| Incentives | Can the organization afford cash incentive programs? |
| Employee Stock Options | Is the organization willing to provide equity based incentives at lower level of the organization? |

Source: Tapamoy, 2009, p.108.

- Compensation management is a mechanism to link employee efforts, performance and outcomes with organizational goals and objectives,
- Organizations need to decide on the compensation level, structure and system for ensuring employee efficiency and cost control.

Compensation level represents an organization's average compensation relative to that of other competing organizations (Gerhart & Milkovich, 1992) and is often labeled as leading, matching or lagging the market. An organization with a policy of leading the market offers higher than the average compensation of the relevant labor market; a policy of lagging the market signifies lower than market average compensation; and policy of matching the market indicates compensation at the relevant labor market average.

Compensation level decisions made by organizations can significantly impact individual and group level consequences, including performance, skill development, work-related attitudes and workforce composition. Compensation level and benefit policy can determine what types of employees are attracted to the organization by signaling job seekers to less visible organizational attributes. These policies will determine whether or not employees are willing to continue to work for the organization. Organizations with higher compensation level should experience increases in both individual and organizational level efficiency because they are more able to attract, retain and motivate the best talents. High compensation levels generate larger applicant pools, which allow organizations to be more selective when hiring and to retain highly qualified employees. High compensation levels may also improve employee and resource efficiency by decreasing employee's unproductive behaviors (Akerlof & Yellen, 1986). If the compensation level or benefit level is not satisfactory, the desirability of movement will increase and it is more likely that withdrawal behaviors such as tardiness, absenteeism, and turnover will increase. Unsatisfactory compensation and benefit levels may also reduce the motivation of employees and eventually negatively affect individual, group and organization performance.

Tapamoy (2009, p.122) says that merit - based pay: reward for productive work have always been considered prime motivation for efforts and achievements. Merit pay is a system wherein employees are evaluated through the instrument of performance appraisal and based on their ratings appropriate raise in fixe compensation (base pay) is given. For example, consider the performance rating plain and related pay increase of an organization as in Table 3.

Thus, if an employee is rated as 'very good', then he shall receive a merit increase of 15% in his base pay, so on and so forth. However, some merit pay schemes may encourage poor work practices as individual employees attempt to maximize their personal gains to the detriment of the entire organization (Hickey & Ichter, 1997). Hammer (1987) cautions that while considering forms of external incentive such as merit pay schemes can fail for a number of reasons including:

- If pay is not related to performance.
- If ratings are seen to be biased.
- If rewards are not viewed as rewarding.

*Table 3. The performance rating plain*

| Performance Rating | Outstanding | Very Good | Good | Average | Marginal |
|---|---|---|---|---|---|
| Pay increase | 20% in base pay | 15% in base pay | 10% in base pay | 5% in base pay | Nil |

Source: Tapamoy, 2009, p.122.

- If there is more emphasis on satisfaction with pay than performance.
- If there is a low level of trust and openness about the merit raises.

The same author add that also exists the knowledge - based pay: Knowledge - based pay has become very popular in new economy organizations and is usually applied to blue-collar work, working in organizations that are in IT services or biotechnology sectors where the work can be specified and defined. Knowledge - based pay seeks to link compensations to depth of the skills, abilities and knowledge a person acquires that are relevant to the work. It compensates employees for knowledge and skills that they possess and the results they produce, not for the job in which they are performing.

Milkovich and Newman (2002) have evaluated the critical processes to determine a knowledge-based structure should include the following steps. An organization must make sure that pay-for-knowledge structure is (Tapamoy, 2009, p.123):

- Internally aligned with work relationships within the organization.
- Perform a skill analysis in order to identify and collect information about knowledge and skills required to perform work in an organization.
- Skill blocks.
- Skill certification.
- The objective of the plan, information required, methods to be used to determine and certify skills, persons to be involved, and usefulness of the results for pay purposes.

Knowledge - based pay systems are considered to be most successful and have been implemented with the greatest ease in organizations with a participative team management style.

Payment by time: According to Heery and Noon (2001), “time rate is a rate of pay based on the amount of time that the employee spends at work”. Under this method (PbT), the compensation is determined on the basis of time worked, which may be hourly, daily, weekly, monthly or any other time period. This system does not take into account the output (e.g., production) of the employees during this time. The wage is calculated as:

Wages = Time spent by the worker x Rate of wage according to time

This system is most suitable when it is not possible to:

- Measure output
- Divide output into units

- Efficiency is important than speed
- Employee is under training

Merits:

- Very simple method of compensation payment.
- Certainty of amount of compensation payment.
- Results is high quality of output, as this method is not tied with quantity of output.
- Better utilization of resources, as this method is not tied with speed.
- This method is most suitable for artistic work.

Another way to compensate human capital is payment by piece rate: Under this method (PbP), employees are paid compensations according to the quantity of output during a given time. The compensation is calculated as (Tapamoy, 2009, p.128):

Compensations = Units of output x rate per unit

This system is most suitable when:

- Work is of standard nature
- Work is measurable
- High pressure to produce more
- Employee is fully trained

Merits:

- Encourages employees to work more.
- Increase in production.
- Efficient utilization of time.
- Less supervision is required.
- Self-justified method.
- Performance is the criteria of compensation payment.
- Decrease in cost of manufacturing.
- Easy and simple to understand and practice.

Another way to compensate talented employees is payment by results: As the competition increases and the conditions become more stiff, organizations recognize that the best results in the shape of outcomes were not obtained under the time-based system and with the object of controlling labour costs, commenced the practice of compensating employees in accordance with the results they produce (Tapamoy, 2009, p.129).

Payment by results is an approach to compensation system management that integrates strategy, people, resources, processes and measurements to improve decision-making, transparency and accountability. The approach focuses on achieving and reporting performance. The aim of Payment by Results (PbR) is to provide a transparent, rules-based system for paying compensation. It rewards efficiency, supports work goals and encourages activity for fostering efficiency. This system ensures a fair and consistent basis for compensation payment rather than being reliant principally on time-based payments.

The system of payment by results has emerged primarily due to the following factors:

- The customers want low prices and quick delivery.
- The organizations want large profits and repeat orders.
- The employees want high compensation and gainful employment.

The system of paying by results is different from paying piece rate method because this system emphasizes not only outputs but outcomes, that is, both the work accomplishments and work behaviours. Results encompass not only outputs but the process, methods and mindset employed for producing the output. For example, a shop floor worker may in his eagerness to increase his compensation based on outputs, increase the speed of the machine above its optimal level thereby damaging the machine in the long run.

Part of the power of such reward comes from knowing that someone took the time to notice the achievement, and find out the responsibilities of the employee. Rewards also provide information to the employees about their effectiveness in various situations (Deci, 1975). Thus, job satisfaction depends upon (Tapamoy, 2009, p.174):

- Intrinsic reward like achievement of the task,
- Extrinsic reward like compensation, promotion, quality of work life programmes, and social relationships,
- Status associated with job level,
- Participative management and empowerment, and
- Work values like fairness, transparency, emotions, dignity and respect in meeting out rewards.

Satisfaction with compensation has been shown to influence overall job satisfaction, motivation and performance, absenteeism and turnover, and may be related to compensation related grievances and lawsuits (Gerhart & Milkovich, 1990; Milkovich & Newman, 2002). Thus, compensation being an extrinsic motivator plays an important role in employee satisfaction. Employees satisfied with their compensation are likely to be productive contributors to organizational goals and objectives and stay in the organization for longer period (Table 4).

The challenge of hiring highly qualified individuals is retaining them. States lost an average of 22 percent of new hires during the probationary period in fiscal year 2006 (GPP 2007). In other words, they retained about 78 percent of the people they hired. Strategic staffing requires that organizations think purposefully about how they intend to retain new hires as well as skilled veterans.

We now analyze the human capital expenses (which comprise the wages of the employees and also the institutions contribution for health and social protection) (Manea M.D., 2012, p.209) from two point of views, such as traditional and the new present one, in order to perceive the strategic role of human capital and the necessity to compensate its performance (Table 5).

The importance of human capital rests with four major issues (Phillips& Pullian Phillips, 2015, p.15):

1. In most, if not all organizations, the cost of employees is the largest single line-item expense. This tremendous cost brings human capital into the focus of executives with two concerns. They want to control the costs, but at the same time, they want to see it as an investment that can reap many returns in the future.
2. Human capital is the last source of competitive advantage. All organizations have relatively the same access to raw materials, financial resources, and technology. The key differentiator to developing a competitive organization is the human

*Table 4. Tangible and intangible costs associated with turnover*

| Financial Costs | Other Costs |
|---|---|
| Decreased productivity | Costs of workflow interruptions when employee leaves |
| Cost of employees who have left organization | Decline in quality of service to citizens |
| Cost of training (time and money) new hires | Loss of expertise and networking connections |
| Indirect costs of recruiting and hiring new employees | Impact on job satisfaction and morale of remaining employees |

Source: Coleman S., 2008.

*Table 5. Human capital perspective*

| Traditional View | Present View |
|---|---|
| • Human capital expenses are considered costs.<br>• The HR functions is perceived as a support staff.<br>• HR is involved in setting the HR budget.<br>• Human capital metrics focus on cost and activities.<br>• Human capital metrics are created and maintained by HR alone.<br>• There is little effort to understand the return on investment in human capital.<br>• Human capital measurement focuses on data at hand.<br>• Human capital measurement is based on what GE and IBM are measuring.<br>• HR programs are initiated without a business need connected to them.<br>• Overall reporting on human capital programs and projects is input focused. | • Human capital expenditures are viewed as a source of value<br>• The HR functions is perceived as a strategic partner.<br>• Top executives are increasingly involved in allocating the HR budget.<br>• Human capital metrics focus on results.<br>• Top executives are involved in the design and use of metrics.<br>• The use of return on investment has become an important tool to understand cause-and-effect relationships.<br>• Human capital measurement focuses on the data needed.<br>• Human capital measurement is based on what is needed in the organization.<br>• HR programs are linked to specific business needs before implementation.<br>• Overall reporting on human capital programs and projects is output focused. |

Source: Phillips& Pullian Phillips, 2015, p.10.

capital, and just like tangible assets from an accounting perspective, which are often called capital assets, the humans in the organization are considered to be important capital. Thus human capital becomes the number-one asset.

3. Human capital is the number-one differentiator in the success of organizations. The most admired companies, the most innovative organizations, the best places to work, and the most successful businesses all attribute their success to how their employees are engaged and drive value for the organization.
4. How human capital is actually employed, just like the employment of financial capital, is the critical issue. In some organizations, the employment of human capital is not very successful, and in others, it is highly successful. The challenge is to make sure that human capital is addressed from the acquisition phase all the way through to optimization.

To understand total rewards we present the content of it (Mathis, Jackson, Valentine& Meglich, 2016, p.404) (Table 6).

Tangible rewards are elements of compensation that can be quantitatively measured, so it is possible to calculate the monetary value of each reward (Mathis, Jackson, Valentine& Meglich, 2016, p.405). Consequently, employees can easily compare the tangible rewards offered by different organizations to determine relative compensation levels. Alternatively, intangible rewards are elements of compensation

*Table 6. Elements of total rewards*

| Tangible Direct Rewards | Tangible Indirect Rewards | Intangible Rewards |
|---|---|---|
| Base Pay<br>• Wages<br>• Salary | Health care Benefits<br>• Medical insurance<br>• Dental insurance<br>• Health spending account | Supportive work environment |
| Variable Pay<br>• Bonuses<br>• Incentives<br>• Equity awards | Paid Time Off<br>• Vacation<br>• Holidays | Challenging work |
| | Pension/retirement benefits<br>• Employee development and training<br>• Education assistance | • Autonomy<br>• Positive reinforcement |

Source: Mathis, Jackson, Valentine& Meglich, 2016, p.404.

that cannot be as easily measured or quantified. How much is working in a positive and supportive office environment worth to an individual? What is the value of being offered enrichment and learning opportunities in the workplace? The perceived value of these intangible rewards can differ among employees, making the development and management of total rewards much more complex. Some employers may find that offering work-life balance opportunities and wellness programs can enhance employees' work satisfaction. Others may find that tangible rewards have a more immediate and positive impact on worker attitudes. This means that total rewards can vary greatly from company to company, and that employers should customize their approach to suit the needs and preference of their workers.

One tangible component of a compensation program is direct compensation, the monetary rewards for work done and performance results achieved. Base pay and variable pay are the most common forms of direct compensation. The most common types of indirect compensation provided to employees are benefits.

## METHODOLOGY

We started our study beginning form some point of views:

- Human capital measurement has been defined by IDS (2004) as being about finding links, correlations and, ideally, causation, between different sets of (HR) data, "using statistical techniques." (Baron & Armstrong, 2007, p.31),
- The CIPD (2004) emphasizes that it deals with the analysis of "the actual experience of employees, rather than stated HR programmes and policies."

- Human capital measurement is based on human capital data comprising the numbers or quantities that describe human resources in an organization, represent particular aspects of the behaviour of people, or indicate the scale of different people management activities.
- There is an overwhelming case for evolving methods of valuing human capital as an aid to decision making. This may mean identifying the key people management drivers and modelling the effect of varying them. The issue is to develop a framework within which reliable information can be collected and analysed such as added value per employee, productivity and measures of employee behaviour (attrition and absenteeism rates, the frequency/severity rate of accidents, and cost savings resulting from suggestion schemes).

## Research Objectives

Our main objective is to analyse the impact could have the accountability (direct costs) of human capital on individual and organizational performance.

## Data Presentation and Analysis

### Variable Description and Establishing the Hypotheses

In this research we used samples of some variables from European level. In this study we used 9 variables (Table 7).

As we may see the earning variable has an increasing trend on the last years, registering a decreasing in the period of economic crisis, so as the number of employees at European level. The administration expenditures are decreasing due to the economic crisis; the people of working age is decreasing and the pensions expenditures are increasing, according to the fact that (ec.europa.eu, EU28):

- Life expectancy at age 65 is increasing from 19.4% (in 2010) to 20% (in 2014),
- The proportion of people of working age in the EU-28 (in 2005 was 67.2% and in 2015 was 65.6%) is shrinking while the relative number of those retired is expanding (in 2005 was 16.6% and in 2015 was 18.9%).

Labor productivity is increasing due to the valuable human capital knowledge, experience and skills and to the new technologies used at work. Labor costs are increasing due to the increasing number of employees as we may see in the table above.

*Table 7. The analyzed variables*

| Obs | Earnings | Employees | Exp. Adm. | Exp. Pens. | Exp. Soc. Ben. | Exp. Soc. Prot. | Income | Lab. Prod. | Lab. Cost |
|---|---|---|---|---|---|---|---|---|---|
| 1 | 15383.94 | 177688.1 | 3.19 | 11.6 | 96.10 | 26.1 | 5.0 | 97 | 19.8 |
| 2 | 15383.94 | 180382.5 | 3.17 | 11.5 | 96.10 | 25.7 | 4.9 | 99 | 19.8 |
| 3 | 15793.36 | 183031.2 | 3.08 | 11.4 | 95.80 | 25.3 | 5.0 | 100 | 19.8 |
| 4 | 15404.52 | 179591.9 | 3.03 | 11.6 | 95.82 | 26.0 | 5.0 | 100 | 21.5 |
| 5 | 14837.37 | 176699.3 | 3.15 | 12.6 | 95.82 | 28.8 | 4.9 | 100 | 21.5 |
| 6 | 15061.46 | 177184.3 | 3.10 | 12.6 | 95.92 | 28.6 | 4.9 | 100 | 21.5 |
| 7 | 15361.28 | 176579.9 | 3.01 | 12.6 | 96.20 | 28.3 | 5.0 | 100 | 21.5 |
| 8 | 16036.40 | 176382.1 | 3.00 | 12.6 | 96.10 | 28.7 | 5.0 | 102 | 23.9 |
| 9 | 16125.33 | 178811.4 | 2.96 | 12.7 | 96.20 | 28.9 | 5.0 | 102 | 24.2 |
| 10 | 16595.42 | 181473.6 | 2.89 | 12.6 | 96.27 | 28.7 | 5.2 | 103 | 24.5 |
| 11 | 17857.28 | 181473.6 | 2.89 | 12.6 | 96.27 | 28.7 | 5.2 | 103 | 25.0 |
| 12 | 17857.28 | 181473.6 | 2.89 | 12.6 | 96.27 | 28.7 | 5.2 | 103 | 25.0 |

Source: www.eurostat.com.

## Hypothesis

For the purpose of this study, the following hypotheses were put to test and analyse them.

**H0:** There is a direct and positive relationship between the analyzed variables:
**H01:** Between the earnings and other analyzed variables,
**H02:** Between the employees and the other analyzed variables.
**H1:** There is a normal distribution among the analysed variables.
**H2:** There is a stationarity among the analysed variables.
**H3:** There is a correlation between the analyzed variables.
**H4:** There is a normal distribution for residuals (skewness = 0 and kurtosis = 3).

## Sample Selection

Data were gathered from European level (28 countries) on 12 years period, between 2005 and 2016.

## Tools Used for Analysis

In this study we were using the following tools:

- Linear Regression Model (to test the impact of two dependent variables on the independent variables),
- Ordinary Least Squares method as simulation methods,
- Descriptive Statistics (to find out the normal distribution of returns of the analysed variables),
- Unit Root Test (to test stationarity among the analysed variables),
- Correlation Matrix (to find correlation between the analysed variables),

The computation of data for this study was made by using E-Views (Version 7.0). The study met some limitations, as follows:

- the study used only 9 variables regarding the analysis of compensation and organization performance,
- the study used a limited period (but important) for analysis- only 12 years from 2005 to 2016,
- and we remind all the limitations associated with the simulation and modeling tools used in this study.

## Using Regression Analysis

Findings of this study show consistency with the theory in the filed, bringing a value in motivation and accountability of human capital and performance obtained through human capital.

According to the literature the model is represented by the following formula, which has been developed in the late eighteen century by Karl Friedrich Gauss and it is described below:

$$y = b_0 + b_1 \times x_1 + b_2 \times x_2 + \ldots + b_n \times x_n \qquad (1)$$

The variables were explained by many authors among them being also Dixon K.R. (2011, p. 125), Darbyshire P., Hampton D. (2014) and Walish P. et al. (2014, p.97), where:

Y = dependent variable,
Xi = independent variables,

$b_1...b_n$ = the slope (or gradient) of the straight line,
$b_0$ = the intercept.

## Using OLS Method

The coefficients of the analysed variables can be estimated by using the OLS method. Further we make a comparative analysis which consists in developing some cases resulting from the combination of independent variables. The results of the simulations processes using Eviews 7 are presented below describing the relationships between variables and their influence on the organization performance.

### First Case: Earnings as a Dependent Variable

In this first case we implemented the Least Squares method, where we used as a dependent variable earnings (Table 8).

The regression model shows that the free term is negative and an increase of 1 point in the earnings will trigger an increase by 0.22 points of the employees, by

*Table 8. Using Least squares where employees is a dependent variable*

| Variable | Coefficient | Std. Error | t-Statistic | Prob. |
|---|---|---|---|---|
| EMPLOYEES | 0.223623 | 0.075848 | 2.948316 | 0.0601 |
| EXPENDITURE_ON_ADMINISTR | 32680.72 | 6877.302 | 4.751969 | 0.0177 |
| EXPENDITURE_ON_PENSIONS | 24178.40 | 5320.182 | 4.544657 | 0.0200 |
| EXPENDITURE_ON_SOCIAL_BE | -2195.904 | 854.6990 | -2.569213 | 0.0825 |
| EXPENDITURE_ON_SOCIAL_PR | -9559.124 | 2082.792 | -4.589572 | 0.0194 |
| INCOME_QUINTILE_SHARE_RA | 11572.68 | 2427.341 | 4.767639 | 0.0175 |
| LABOR_W_PERSON_EMPLOYEED | -202.8226 | 193.9304 | -1.045852 | 0.3725 |
| LABOUR_COSTS_ANNUAL_DATA | 2153.303 | 423.5639 | 5.083773 | 0.0147 |
| C | -29288.71 | 60371.00 | -0.485145 | 0.6608 |
| R-squared | 0.989426 | Mean dependent var | | 15974.80 |
| Adjusted R-squared | 0.961228 | S.D. dependent var | | 1002.714 |
| S.E. of regression | 197.4402 | Akaike info criterion | | 13.52245 |
| Sum squared resid | 116947.9 | Schwarz criterion | | 13.88613 |
| Log likelihood | -72.13472 | Hannan-Quinn criter. | | 13.38781 |
| F-statistic | 35.08888 | Durbin-Watson stat | | 2.109917 |
| Prob(F-statistic) | 0.007001 | | | |

Sample: 1 12, Included observations: 12

32,680.7 points of the expenditures on administration, by 24,178.4 of the expenditures on pensions, by 11,572.6 of the income and by 2,153.3 of labor costs and a decrease by 2.195.9 of social benefits, by 9,559.1 of expenditure on social pr, and by 202.8 of labor productivity.

R-squared is 0.989, resulting that R is 0,994, concluding that between earnings and the analyzed variables there is a strong, direct and positive relationships (being closed to 1). Thus, the hypothesis H01- There is a relationship between earnings and the other analyzed variables is accepted.

Standard error of the regression is very high, indicating that the forecasting error variance is very large, and the model, although valid, explains the dependence between the *dependent variable* and the other independent variables taken into consideration.

In order to validate the model the results of the Fisher - Snedecor, Durbin Watson and White statistic tests are interpreted and for the analysis regarding the normality of residuals, Jarque Bera test is used. F statistic test has a value of 35.08, higher than the table value of 2.21, and the probability of 0.007 is very small for a completely new research such as that undertaken in this article. All the tests performed confirm that the assumption of homogeneity of the residual variable is met.

### Second Case: Employees as a Dependent Variable

We take in the second case as dependent variable the employees and we applicated the Laest Square method, obtaining the following calculation (Table 9).

Analyzing this calculation, we observe that F-stat is very high again and probability is quit low.

R-squared (0.97) is showing again a very strong and positive relationship between the analyzed variables, resulting that R is 0.98 (being very close to 1), concluding that between employees and the analyzed variables there is a strong, direct and positive relationships. Thus, the hypothesis H02- There is a relationship between employees and the other analyzed variables is accepted.

## Descriptive Statistics

Table 10 presents a summary of the descriptive statistics using data between 2005-2016.

Table 9 shows the results of descriptive statistics for the analysed variables during the study period from 2005 to 2016. We may observe that in the summary statistics were used mean, minimum, maximum, median, standard deviation (SD), skewness, kurtosis and the Jarque- Bera in order to analyse the sample during the period noted above. Thus, the description of data series distributions under general trend, values diversity and form is shown in table above.

*Table 9. Using Least squares where employees is a dependent variable*

| Variable | Coefficient | Std. Error | t-Statistic | Prob. |
|---|---|---|---|---|
| EXPENDITURE_ON_ADMINISTR | -112108.1 | 42500.92 | -2.637780 | 0.0778 |
| EXPENDITURE_ON_PENSIONS | -85685.70 | 29502.25 | -2.904378 | 0.0623 |
| EXPENDITURE_ON_SOCIAL_BE | 8170.744 | 3535.768 | 2.310882 | 0.1039 |
| EXPENDITURE_ON_SOCIAL_PR | 33592.17 | 11881.15 | 2.827349 | 0.0663 |
| INCOME_QUINTILE_SHARE_RA | -36569.89 | 17477.95 | -2.092344 | 0.1275 |
| LABOR_W_PERSON_EMPLOYEED | 1185.896 | 542.3851 | 2.186446 | 0.1166 |
| LABOUR_COSTS_ANNUAL_DATA | -7772.536 | 2346.657 | -3.312175 | 0.0453 |
| EARNINGS | 3.324469 | 1.127582 | 2.948316 | 0.0601 |
| C | 37564.99 | 240756.3 | 0.156029 | 0.8859 |
| R-squared | 0.970674 | Mean dependent var | | 179231.0 |
| Adjusted R-squared | 0.892471 | S.D. dependent var | | 2321.542 |
| S.E. of regression | 761.2699 | Akaike info criterion | | 16.22156 |
| Sum squared resid | 1738596. | Schwarz criterion | | 16.58524 |
| Log likelihood | -88.32935 | Hannan-Quinn criter. | | 16.08691 |
| F-statistic | 12.41229 | Durbin-Watson stat | | 2.597893 |
| Prob(F-statistic) | 0.031254 | | | |

Sample: 1 12, Included observations: 12

*Table 10. The results of descriptive statistics for analysed variables during 2005 and 2016*

| Obs | Earn | Empl | Exp. Adm. | Exp. Pens. | Exp. Soc. Ben. | Exp. Soc. Prot. | Income | Lab. Prod. | Lab. Cost |
|---|---|---|---|---|---|---|---|---|---|
| Mean | 15974.8 | 179231.0 | 3.03 | 12.25 | 96.07 | 27.70 | 5.02 | 100.75 | 22.33 |
| Median | 15598,9 | 179201.7 | 3.02 | 12.60 | 96.10 | 28.65 | 5.00 | 100.00 | 21.50 |
| Maximum | 17857.2 | 183031.2 | 3.19 | 12.70 | 96.27 | 28.90 | 5.20 | 103.00 | 25.00 |
| Minimum | 14837,3 | 176382.1 | 2.89 | 11.40 | 95.80 | 25.30 | 4.90 | 97.00 | 19.80 |
| Std. dev. | 1002.7 | 2321.54 | 0.10 | 0.53 | 0.18 | 1.44 | 0.11 | 1.86 | 2.06 |
| Skewness | 0.99 | 0.15 | 0.03 | -0.72 | -0.42 | -0.73 | 0.65 | -0.32 | 0.079 |
| Kurtosis | 2.759 | 1.59 | 1.69 | 1.58 | 1.62 | 1.65 | 2.09 | 2.38 | 1.45 |
| Jarque-Bera | 1.99 | 1.03 | 0.85 | 2.05 | 1.30 | 1.97 | 1.25 | 0.40 | 1.20 |
| Probability | 0.369 | 0.594 | 0.65 | 0.35 | 0.51 | 0.37 | 0.53 | 0.81 | 0.54 |
| Sum | 191697.6 | 2150772 | 36.36 | 147.0 | 1152.87 | 332.50 | 60.30 | 1209.0 | 268.0 |
| Sum Sq. Dev | 11059800 | 59285138 | 0.13 | 3.19 | 0.37 | 23.02 | 0.14 | 38.25 | 46.88 |
| Observations | 12 | 12 | 12 | 12 | 12 | 12 | 12 | 12 | 12 |

Author's calculation using E-views 7 version.

All the analyzed variables present a positive mean values. The mean for employees and for earnings is very high. Between the expenditures, the high mean value is for expenditures with pensions.

The range of variation between maximum and minimum is quite logical. We observe in the table above that for all data series, the mean and median have very similar value, the ratio between mean and the median of each variable being approximately 1.

The standard deviation compared to the mean is not low which indicates coefficient of variation. Also, the sum squared deviation row represents the net change over the sample period. It shows that number of employees and their earnings increased very much, while expenditures of all the analysed variables, income increase insignificantly, have a medium increase. The employees variable has the largest standard deviation among all the other analyzed variables. Also, the standard deviation (Std. Dev.) of data series has small values for all the variables except the earnings and the employees and therefore it can be considered that the series are relatively homogeneous. The variables with higher values for standard deviations, can be explained by the wide range of values for these series due to the economical development and increase from the period of 2007-2012 and increase of number of employees in the period 2007-2008 and 2014-2016 for some countries from all the 28 European countries, data included in the model.

In terms of skewness, the results for variable 1,2,3,7 and 9 are positively skewed while for variable 4,5,6 and 8 indicate a negative skewness.

The value for kurtosis in each variable is between 1.75 and 2.75 being below the benchmark for a normal distribution of 3, which is positioned near normality. It is important to show that all the values of kurtosis are smaller than three, but bigger than 0, making the distribution Leptokurtic and the values concentrated around the central tendency. Thus, the analyzed variables are characterized by a normal distribution.

The values obtained for Jarque Bera test are between 0,40 and 2.05 (indicating that all the variables are approximately normally distributed), having associated probabilities closed to zero indicating that the variables are not so volatile. This can be explained by the influences of external factors such as compensation policies, cultural, social, political or national differences in the European Union. The Jarque-Bera (JB) values show that all the sample indices were normally distributed. In other words, all the sample indices were less volatile during the analysed period. Thus, the distribution of all the sample indices was normal.

So, the hypothesis H1- There is a normal distribution among the analysed variables, was accepted.

## Using Unit Root Test (URT)

Panel data methodology is used in this research, combing cross-section and time series data. Thus, we must verify the stationary level of the analysed data, using Unit Root test. We used in this study E-views 7 implementing three unit root tests (Augmented Dickey-Fuller, DF-GLS, and Phillips-Perron) in order to determine the stationarity of the analysed variables (Table 11).

Dickey and Fuller (1979) show that under the null hypothesis of a unit root, this statistic does not follow the conventional Student's t-distribution, and they derive asymptotic results and simulate critical values for various test and sample sizes. More recently, MacKinnon (1991, 1996) implements a much larger set of simulations than those tabulated by Dickey and Fuller. As we may observe:

- **For Earnings:** ADF statistic value is -1.04 and the associated one-sided p-value (for a test with 11 observations) is 0.889; the statistic value (-1.04) is greater than the critical values (-5.12, -3.93, and -3.42) so that we do not reject the null hypothesis at conventional test sizes.
- **For Employees:** ADF statistic value is -2.16 and the associated p-value (for a test with 10 observations) is 0.45; the statistic value is greater again than the critical values, so the hypothesis is accepted.
- **For Expenditures on Administration:** ADF statistic value has a p-value of 0.17; statistic value is greater than the critical values, so the hypothesis is accepted.
- **For Expenditures on Pensions:** ADF has a p-value equal to 0.81; statistic value is greater than the critical value,
- For social benefits, for social protection, income, labor productivity and also for labor costs, their associated statistic values are greater than their critical values, so the hypothesis is rejected and we may conclude that there is stationarity among the analysed variables.

The results of Unit Root test indicate the fact that the returns data of all sample obtained stationarity during the study period. Hence the hypothesis H2- There is a stationarity among the analyzed variables, is accepted.

These unit root tests determine whether the variables are stationary or not and also indicates the degree of integration.

## The Correlations Among the Analysed Variables

A correlation is a statistic function, simple but it can explains if there is any relationship between two analyzed variables

*Table 11. Critical values at 1%, 5% and 10% using URT for the analysed variables (2005-2016)*

| Variable/Unit Root Test | | ADF Test | | | DF-GLS Test | | PP Test | | |
|---|---|---|---|---|---|---|---|---|---|
| | | Stat Value | Crit Value | P Value/Nb of Observations | Stat Value | Crit Value | Stat Value | Crit Value | P Value |
| Earnings | 1% | | -5.12 | | | -2.79 | | -4.20 | |
| | 5% | -1.04 | -3.93 | 0.889/11 | -0.027 | -1.97 | 0.31 | -3.17 | 0.96 |
| | 10% | | -3.42 | | | -1.60 | | -2.72 | |
| Employees | 1% | | -5.29 | | | -2.81 | | -2.79 | |
| | 5% | -2.16 | -4.008 | 0.45/10 | -2.48 | -1.98 | 0.52 | -1.97 | 0.81 |
| | 10% | | -3.46 | | | -1.60 | | -1.60 | |
| Exp. on admin | 1% | | -5.29 | | | -2.79 | | -2.79 | |
| | 5% | -3.03 | -4.008 | 0.17/10 | -1.08 | -1.97 | -4.18 | -1.97 | 0.0007 |
| | 10% | | -3.46 | | | -1.60 | | -1.60 | |
| Exp. on pensions | 1% | | -5.12 | | | -2.79 | | -2.79 | |
| | 5% | -1.35 | -3.93 | 0.81/11 | -1.21 | -1.97 | 0.92 | -1.97 | 0.89 |
| | 10% | | -3.42 | | | -1.60 | | -1.60 | |
| Exp. on social ben | 1% | | -5.12 | | | -2.79 | | -2.79 | |
| | 5% | -2.29 | -3.93 | 0.40/11 | -1.06 | -1.97 | 0.35 | -1.97 | 0.77 |
| | 10% | | -3.42 | | | -1.60 | | -1.60 | |
| Exp. on soc prot | 1% | | -5.12 | | | -2.79 | | -2.79 | |
| | 5% | -1.59 | -3.93 | 0.72/11 | -1.22 | -1.97 | 1.41 | -1.97 | 0.94 |
| | 10% | | -3.42 | | | -1.60 | | -1.60 | |
| Income | 1% | | -5.52 | | | -2.79 | | -2.79 | |
| | 5% | -0.16 | -4.10 | 0.97/9 | -0.90 | -1.97 | 1.22 | -1.97 | 0.93 |
| | 10% | | -3.51 | | | -1.60 | | -1.60 | |
| Labor prod/pers. Empl | 1% | | -5.52 | | | -2.79 | | -2.79 | |
| | 5% | -2.78 | -4.10 | 0.23/9 | -1.16 | -1.97 | 2.24 | -1.97 | 0.98 |
| | 10% | | -3.51 | | | -1.60 | | -1.60 | |
| Labor costs | 1% | | -5.12 | | | -2.79 | | -2.79 | |
| | 5% | -3.00 | -3.93 | 0.17/11 | -0.63 | -1.97 | 5.26 | -1.97 | 0.99 |
| | 10% | | -3.42 | | | -1.60 | | -1.60 | |

Calculation made by the authors using E-views 7.0.

In Table 12 we used the correlation matrix to shows the results of this correlation. According to the results, the values of correlation ranged from (-0.925) between labor productivity and expenditure on administration to (0.995) between expenditures on social protection and expenditures on pensions.

*Table 12. The correlation between the analyzed variables*

| Correl | Earn. | Empl. | Exp. Adm. | Exp. Pens. | Exp. Soc. Ben | Exp. Soc. Prot | Income | Lab. Prod | Lab. Costs |
|---|---|---|---|---|---|---|---|---|---|
| **Earn** | 1 | | | | | | | | |
| **Empl** | 0.59 | 1 | | | | | | | |
| **Exp. Adm** | -0.82 | -0.39 | 1 | | | | | | |
| **Exp. Pens** | 0.34 | -0.34 | -0.58 | 1 | | | | | |
| **Exp. Soc. Ben** | 0.68 | 0.14 | -0.63 | 0.48 | 1 | | | | |
| **Exp. Soc. Prot** | 0.35 | -0.33 | -0.57 | 0.99 | 0.48 | 1 | | | |
| **Income** | 0.91 | 0.58 | -0.85 | 0.31 | 0.67 | 0.32 | 1 | | |
| **Lab. Prod** | 0.77 | 0.35 | -0.92 | 0.67 | 0.52 | 0.67 | 0.71 | 1 | |
| **Lab. Costs** | 0.78 | 0.16 | -0.90 | 0.76 | 0.65 | 0.78 | 0.74 | 0.93 | 1 |

The following calculations shows the correlations among the analyzed variables. The value obtained in Table 12 shows us the strength of the relationship between two-by-two analyzed variables.

- If the value is a perfect zero, that means that the two variables are not related to each other at all,
- If the value is between zero to one, it means that the relationship becomes stronger and stronger, and if it is closer to one, means that the two variables are very strongly related to each other.
- If the value is between zero and minus one, it means that we have a low correlation and the two variables are a little bit related to each other, or not at all.

We can see that:

- **Earnings:**
  - **Have a Positive Relationship With:** Income (0.909), labor costs (0.78) and labor productivity (0.77) and are positive and strong relationships; number of employees (0.597) and exp. on soc ben (0.68) and are positive

but not so stronger relationships, exp. on pensions (0.34) and exp. on soc prot (0.35) there are positive but wick correlations.
    - **Have a Negative Relationship With:** Expenditure on administration (-0.82) which is a negative and strong relationships, which means that the two variables move in the opposite direction from each other.
- **Employees:**
    - **Have a Positive Relationship With:** Exp. on soc ben (0.14) and with labor costs (0.16) but wick and with income (0.58) and with labor productivity (0.35) being stronger relationships,
    - **Have a Negative Relationship With:** Exp. on adm (-0.39), exp on pensions (-0.34), exp. on soc prot (-0.33), but are not so stronger.
- **Expenditure on Administration:**
    - **Have Relationships With:** All the rest analyzed variables only negative strong relationships situated among (-0.92 and -0.57).
- The rest of the variables have only positive relationships starting with the highest (0.99 between expenditure on pensions and expenditure on social protection) to the wickest (0.31 between expenditure on pensions and income).

From these calculations results that the hypothesis H3- There is a corelation between the analyzed variables, was accepted, thus between the analyzed variables do exist more direct and positive correlations, with a few exceptions presented above.

## The Normality for Residuals

To test the residuals of the model, we have to analyze if they follow a normal distribution or not.

To verify the hypothesis of normality for residuals, we calculate the Jarque-Bera test, the skewness, which measure the symmetry of the residuals distribution around the mean of these (and is zero), and the kurtosis. The Jarque-Berra test (JB) is based on the hypothesis that the normal distribution has the skewness, $S = 0$, and the kurtosis, $K = 3$.

If the probability p of JB is enough low, then the normality hypothesis of residuals is rejected, and if the probability is high, the normality of residuals distribution is accepted. In our case (Table 13) the value of JB test is 0.76. It is observed that skewness = 0.34, and kurtosis =1.97, the probability of the test is =0.68. From these reasons we accept the hypothesis that the distribution of residuals follows an almost normal distribution. Thus, the hypothesis *H4- There is a normal distribution for residuals* is partial accepted. This is observed also from the representation made by Eviews, which shows that the distribution of residuals is almost normal.

*Table 13. Jarque-Bera test for residuals*

| Series | RESID |
|---|---|
| Sample | 12 |
| Observations | 12 |
| Mean | 6.90e- 15 |
| Median | -0.001013 |
| Maximum | 0.006714 |
| Minimum | -0.005509 |
| Std.Dev. | 0.003985 |
| Skewness | 0.340827 |
| Kurtosis | 1.971205 |
| Jarque-Bera | 0.761536 |
| Probabilty | 0.683336 |

Source: Data from Eviews 7.0. calculations.

# RESULTS, DISCUSSIONS, AND RECOMMENDATIONS

## Results and Discussions

Regarding the research hypothesis, according to the calculations made, we may conclude that:

**H0:** There is a direct and positive relationship between the analyzed variables:

**H01:** Between the earnings and other analyzed variables. R-squared is 0.989, resulting that R is 0,994, concluding that between earnings and the analyzed variables there is a strong, direct and positive relationships (being closed to 1). Thus, the hypothesis is accepted.

**H02:** Between the employees and the other analyzed variables. R-squared (0.97) is showing again a very strong and positive relationship between the analyzed variables, resulting that R is 0.98 (being very close to 1), concluding that between employees and the analyzed variables there is a strong, direct and positive relationships.

**H1:** There is a normal distribution among the analysed variables. In terms of skewness, the results for variable 1,2,3,7 and 9 are positively skewed while for variable 4,5,6 and 8 indicate a negative skewness. The value for kurtosis in each variable is between 1.75 and 2.75 being below the benchmark for a normal distribution of 3, which is positioned near normality. It is important to show that all the values of kurtosis are smaller than three, but bigger than

0, making the distribution Leptokurtic and the values concentrated around the central tendency. Thus, the analyzed variables are characterized by a normal distribution. The values obtained for Jarque Bera test are between 0,40 and 2.05 (indicating that all the variables are approximately normally distributed), having associated probabilities closed to zero indicating that the variables are not so volatile. From all these calculation results that H1 is accepted.

**H2:** There is a stationarity among the analysed variables. The results of Unit Root test indicate the fact that the returns data of all sample obtained stationarity during the study period. Hence the hypothesis H2- There is a stationarity among the analyzed variables, is accepted.

**H3:** There is a correlation between the analyzed variables. Analyzing the relationships between the variables, and from these calculations results that the hypothesis H3- There is a corelation between the analyzed variables, was accepted, thus between the analyzed variables do exist more direct and positive correlations, with a few exceptions presented above.

**H4:** There is a normal distribution for residuals (skewness = 0 and kurtosis = 3). the value of JB test is 0.76. It is observed that skewness = 0.34, and kurtosis =1.97, the probability of the test is =0.68. From these reasons we accept the hypothesis that the distribution of residuals follows an almost normal distribution. Thus, the hypothesis H4- There is a normal distribution for residuals is partial accepted.

As a final conclusion, we may observe that there are positive and strong relations between the human capital (employees variable) and their costs. Analyzing the role in obtaining performance, no matter how large are the sums invested in talented employees, but the results are important and by using the business simulation (which consists in using simulation and Eviews) the organization has more performance by gaining new knowledge, skills, experience and implicitly organizational performance through financial motivation.

## RECOMMENDATIONS, CONCUSION, AND FUTURE WORK DIRECTIONS

Due to the relative low costs for collecting information necessary for making decisions, simulation and modeling are used in domains such as manufacturing (Robinson, 2004, p.11), public and social field, transportation, construction, or telecommunication (Fishman, 2001, p.3), reengineering, IT, gaming, management, human and social system (Barjis & Pergl, 2014, p.100), marketing (Baker, 2001, p.21), finance (Davis & Pecar, 2013, p.551), science, arts, music and mathematics

(Chang, 2011, p.1). Simulation is also applied in human resources field such as training and learning (Anton et al., 1999, p.7), development, social dysfunctional analysis (Florea & Mihai, 2016). We may see the advantages of using simulation techniques in human capital field, so the organizations if they will implement such effective technics they will predict future risks and problems and they will obtain performance through collaboration, relationships, team work, gaining new skills and new knowledge. Now, when IT organizations develop continuously new simulation programs and learning to simulate being not so hard, specialists and organizations must make investments in developing e-skills in order to obtain individual and organizational performance.

The study set out to carry out an empirical analysis on the relationship between human capital and human capital expenses in order to increase performance. This study affirms that the time series data employed in this study are using different tests to determine the important role of human capital in obtaining performance, by using financial motivation (expenditures on administration, on pensions, on social benefits or on social protection). Having motivated and satisfied employees the organizations may achieve its objectives by obtaining competitive advantage.

## REFERENCES

Akay, V. (2006). *Business is the people and people are the business: Humanizing the corporation*. iUniverse.

Altiok, T., & Melamed, B. (2000). *Simulation modeling and analysis with ARENA*. Elsevier.

Anderson, P. H., Beveridge, D. A., & Scott, T. W. (2000). *Threshold Entrepreneur. A New Business Simulation*. Upper Saddle River, NJ: Prentice-Hall Inc.

Anton, J. (1999). *Call center performance enhancement. Using simulation and modeling*. Purdue University Press.

Armstrong, M., & Baron, A. (2003). *Strategic HRM the key to improve business performance*. CIPD.

Baker, M. J. (2001). *Marketing. Critical perspectives on business and management*. Routledge.

Banks, J. (1996). *Discrete-Event System Simulation*. Englewood Cliffs, NJ: Prentice-Hall.

Barjis, J., & Pergl, R. (2014). Enterprise and organizational modeling and simulation. *10th In. Workshop, EOMAS 2014.*

Barney, J. (1991). Firm resources and sustained competitive advantage. *Journal of Management, 17*(1), 99–120. doi:10.1177/014920639101700108

Baron, A., & Armstrong, M. (2007). *Human capital management.* Kogan Page Pub.

Bashner, D. B. (2014). *Employee benefits design and compensation.* FT Press.

Bohlander, G., & Snell, S. (2010). *Managing human resources.* Cengage Learning.

Boxall, P. F., & Purcell, J. (2007). *The Oxford handbook of HRM.* Oxford University Press.

Brewster, C., & Larsen, H. H. (2007). *Human resources management in Northern Europe. Trends, dilemas and strategies.* Blackwell Business.

Bridgeland, D. M., & Zahavi, R. (2008). *Business modeling: a practical guide to realizing business value.* Morgan Kaufmann.

Chang, M. (2011). *Monte Carlo Simulation for the Pharmaceutical industry. Concepts, algorithms and care studies.* CRC Press.

Coleman Selden, S. (2008). *Human capital.* CQ Press.

Davis, G., & Pecar, B. (2013). *Quantitative methods for decision making using Excel.* Oxford University Press.

DeCenzo, D. A. (1988). *Personnel/Human resouce management.* Englewood Cliffs, NJ: Prentice-Hall.

Duica, A. (2016). Using simulation and six sigma to analyze and improve motivation process of teachers from undergraduate institutions. *VJES, 0*(0), 94–103.

Faulkner, D. (2002). *Strategy: critical perspectives on business and management.* Routledge.

Fishman, G. S. (2001). *Discrete-event simulation: modeling, programming, and analysis.* Springer Science&Business Media. doi:10.1007/978-1-4757-3552-9

Fishman, G. S. (2013). *Discrete-Event Simulation: Modeling, Programming, and Analysis.* Springer Science&Business Media.

Florea, N. V. (2015). *Improving employees and organization performance using business simulation and modeling. In Management & Marketing* (pp. 367–381). University of Craiova.

Florea, N. V. (2017). *Simulare si modelare in afaceri. Studii de caz.* Bucuresti: Editura Mustang.

Florea, N. V., & Mihai, D. C. (2015). Improving organization performance through human capital development using a regression function and MATLAB. *JOSA, 3*(32), 229–238.

Florea, N. V., & Mihai, D. C. (2017). Predicting employees evaluation performance using simulation and mathematical modeling. *JOSA, 1*(38), 81–94.

Harris, M. M. (2012). *Handbook of research of international HRM.* Psychology Press.

Holbeche, L., & Mayo, A. (2012). *Motivating people in Lean organizations.* Butterworth-Heineman.

Janssen, M. (2015). *Policy Practice and Digital Science: Integrating Complex Systems, Social Simulation and Public Administration in Policy Research.* Springer. doi:10.1007/978-3-319-12784-2

Laguna, M., & Marklund, J. (2011). *Business process modeling, simulation and design.* Pearson Edu.

Lakhdar, S. (2001). *Gestion de ressources humaines.* Universite de Boeck.

Law, A. M., & Kelton, D. W. (1991). *Simulation modeling and analysis.* McGraw-Hill.

Lawal, B. (2014). *Applied Statistical Methods in Agriculture, Health and Life Sciences.* Springer.

Manea, M. D. (2012). *Contabilitatea instituțiilor publice.* Bucuresti: Academic Press.

Manolescu, A. (2006). *Managementul resurselor umane.* Bucuresti: Ed. Economica.

Mathis, R., Jackson, J., Valentine, S., & Meglich, P. (2016). *Human resources management.* Cengage Learning.

McHaney, R. (2009). *Understanding computer simulation.* Ventus Pub., ApS.

Millmore, M. (2007). *Strategic HRM: contemporary issues.* Pearson Education.

Mittal, S. (2017). *Guide to Simulation-Based Disciplines: Advancing Our Computational Future.* Springer. doi:10.1007/978-3-319-61264-5

Moore, J. H., & Weatherford, L. R. (2001). *Decision Modeling with Microsoft Excel.* Prentice-Hall.

Moynihan, D. P., & Pandey, S. K. (2007). The role of organizations in fostering public service motivation. *Public Administration Review*, *67*(1), 40–53. doi:10.1111/j.1540-6210.2006.00695.x

Palmer, S. (2012). *People an self management*. Butterworth-Heineman.

Phillips, J., & Pullian Phillips, P. (2015). *High impact human capital startegy*. AMACOM.

Powell, G. R. (2008). *Marketing calculator. Measuring and managing return on marketing investment*. Singapore: John Wiley & Sons.

Prasad, A. (2011). *Future Internet Services and Service Architectures*. River Pub.

Reece, B. L. (2011). *Effective human relations. Interpersonal and organizational applications*. South-Western Cengage Learning.

Reed, A. (2001). *Innovation in HRM: tooling up for the talent wars*. London: CIPD House.

Robinson, S. (2004). *Simulation. The practice of model development and use*. John Wiley & Sons.

Rodriguez, J. (2012). *Trends in Practical Applications of Agents and Multiagent Systems*. Springer. doi:10.1007/978-3-642-28795-4

Sa Silva, P. (2011). *Ch.* Simulation, Games and challenges from schools to enterprises. In *Enterprise and organizational modelling and simulation, 7th International Workshop, EOMAS 2011*. Springer.

Sabuncuoglu, I. (2015). *Industrial engineering applications in emerging countries*. CRC Press.

Seddighi, H. (2000). *Econometrics: A Practical Approach*. Psychology Press.

Seddighi, H. (2013). *Introductory Econometrics: A Practical Approach*. Routledge.

Sims, R. R. (2002). *Organizational success through effective human resources management*. Greenwood Pub.

Sokolowski, J. A., & Banks, C. M. (2011). *Principles of modeling and simulation. A multidisciplinary approach*. John Wiley & Sons.

Squazzoni, F. (2009). *Epistemological Aspects of Computer Simulation in the Social Sciences*. Springer. doi:10.1007/978-3-642-01109-2

Storey, J. (2009). *The Routledge companion to SHRM*. Taylor&Francis.

Tanasescu, D. A. (2014). *E-Communication and E-development, efficient response for customer expectations into an E-economy environment*. ECR.

Tapamoy, D. (2009). *Compensation management*. Excel Books.

Taylor, S. (2014). *Agent-based Modeling and Simulation*. Springer. doi:10.1057/9781137453648

Wainer, G., & Mosterman, P. (2016). *Discrete-Event Modeling and Simulation: Theory and Applications*. CRC Press.

## KEY TERMS AND DEFINITIONS

**Human Capital:** That unique resource and strategic partner which through its tangible assets such as skills, expertise, abilities, competencies bring individual and organization performance, competitive advantage on the market and achieve objective.

**Human Capital Expenses:** Important costs for organizations but now perceived as future investments due to its uniqueness and its ability to obtain long term performance.

**Modeling:** It is necessary when the direct experiencing on the real system is not possible or recommended, building a model which must be simple, reflect the reality, be intuitive but rigorous, simple, efficient, reliable, verifiable, documented, qualitative, available, complete, robust, easy to control, adapt, work with it, and modified.

**OLS Method:** A discrete event simulation and a practical method for checking properties of the estimators.

**Simulation Techniques:** A necessary process which consisting in detection of those future actions and creating of a future environment wanted in order to collect the data and information regarding the way of behaviour of the analysed process from the real life without any prior testing.

# Chapter 5
# Improve Center of Sterilized Material Performance:
## A Discrete Event Simulation Model

**Daniel Bouzon Nagem Assad**
*Rio de Janeiro State University, Brazil*

**Thais Spiegel**
*Rio de Janeiro State University, Brazil*

**Ana Carolina Pereira de Vasconcelos Silva**
*Rio de Janeiro State University, Brazil*

## ABSTRACT

*The health sector in Brazil faces difficulties related mainly to financial, administrative, and organizational structure. Center of Sterilized Material (CSM) provide material for the other components of the hospital, and is essential for the proper working of medical care. This work investigates CSM strategy operation in a public hospital. There is an expectation that it needs additional human resources to meet the current demand. During the research, we evaluated this increment and showed that the critical resource is not the human one. We went forward and presented an alternative that does not prescind of investment and increase the installed capacity. The contributions of this research, brings, in the midst of the financial crisis experienced, the possibility of meeting more and better the internal demand, without any financial contribution. It's about showing the potentiality of bringing together professionals and academics, applying methods and management tools in the health sector.*

DOI: 10.4018/978-1-5225-3226-2.ch005

# INTRODUCTION

## Context: Management of Hospital Operations

In the last few decades, hospitals across the world have faced increased economic pressure (CAPKUN, et al., 2012). Modernization of hospital structures has a two-fold objective: Improving quality of care and reducing healthcare costs (REYMONDON et al., 2008). The increasing demand for productivity and quality in service production was recognized 20 years ago yet the issue remains current (SUNDBO, 1994; VAHATLO & KALLIO, 2015).

The health sector in Brazil faces difficulties related mainly to financial, administrative problems as well as the ones in the organizational structure (MALAGÓN-LONDOÑO et al., 2003); and it's marked by the tendency of increasing costs in the assistance, enhanced by the increase aging of the population. The inefficiencies deriving from the waste in the health service management constitute a problem (ARAUJO, 2005). One searches, however, to obtain efficiency gains in health services, reconciling different dimensions, such as quality, reliability, speed, availability with efficiency in operations and cost compatibility (SPIEGEL & ASSAD, 2016).

Within the service operations strategy literature, there is a small but growing body of work that explores the links between operations strategy and healthcare performance (SILVA et al., 2015). As discussed above, with healthcare costs rising, the increasing importance of quality in healthcare, and current demographic patterns, this link is now even more important. Operations strategy has flourished as a field, yet there is still much to be learned regarding how this knowledge base can be effectively applied within the healthcare setting. Specifically, how transferable are these manufacturing-derived principles to a setting where quality and costs take on a very different meaning (SPIEGEL & CAMEIRA, 2016)?

## CSM

Within health facilities, any medical supply that is foreseen to be in contact with a patient must be sterilized (PERKINS, 1983). Sterile supplies or instruments (i.e., sterile devices) are either designed for single use or for re-use. Hospital sterilization is now organized in order to implement a quality assurance system(PITTET et al., 2008). The hospital sterilization is responsible for supplying, preparing, distributing and managing the sterile medical devices. Centralizing these activities, allowing a grouping of means and competencies, leads to benefits in health security and economics in the short and medium-term (SEAVEY et al., 2007; AORN, 2012).

The Center of Sterilized Material (CSM) is a standalone unit. As the definition of RDC nº 50 regulatory current in Brazil, the CSM is a critical area, meaning, an environment where there is an increased risk of transmitting infection, where risk procedures are being held without patients (BRAZIL, 2002). It's considered a sector of middle activity, since it serves customers and suppliers, providing service to the entire complex of the hospital. It receives inputs from other service providers for its activities, as packaging, chemical tests, disinfectants and surgical instruments (PEZZI & LEITE, 2010; POSSARI, 2010; SOBECC, 2013).

Accordingly, study the CSM is relevant, because it is an area of great importance for hospitals since its operational performance contributes in a significant way to reach the increase of the quality of the offered service and the cost reduction (REYMONDON et al., 2008).

## Contributions for This Work

Within this context, the work investigates CSM strategy operation in a reference public hospital. The investigation is basically conduced in three steps: process description, measurement of each phase and estimate the demand based on historical data. We aim to understand the system behavior, the bottlenecks resources and its results terms of, waiting and total time of each entity using discrete event simulation. According to Günal & Pidd (2010) literature review, discrete event simulation (DES) has been widely used in modeling health-care systems in many different units like operation room, critical care units, laboratory and pharmacy, but it was found no research involving CSM and DES.

For instance, many literature reviews has been conducted showing the importance of considering other process upstream and downstream operating rooms like Cardoen et al. (2010), Guerriero & Guido (2011), Demeulemeester et al. (2013), but none research involving CSM. The same result can be seen in Hulshof et al. (2012) where the aim was to propose a taxonomic classification of planning decisions in health care where CSM was not mentioned.

Finally, Günal & Pidd (2010) concluded that even after 25 years of Wilson (1981) review, There are still many barriers to the successful implementation of simulation to some degree in all domains, including health care and expose reasons:

1. A simulation project is often initiated by decision-makers who seek urgent solutions to their particular problems, having already carried out a thorough analysis of the situation. As the timing is crucial for the decision-makers, simulation analysts are expected to generate quick solutions, which they failed to do so, due to the time spent in collecting and analyzing data.

2. The simulation process is expensive (this may have changed due to better software).
3. When quick solutions are expected, modelers tend to oversimplify the models, which can cause decision-makers lack of confidence.

Therefore, in this model our goal is to reduce the total time of surgical entities to permit instruments re-use at the operation theater on the same day. Thus, we propose some scenarios increasing professionals and increasing bottleneck capacity, in other words, increasing costs, but, at least, we concluded that it's possible to reduce the total time of surgical entities and improve the system efficiency proposing alternative materials arrival rules. We reduced the average waiting time and total time of surgical entities in almost 30%, without increasing costs. We analyze and quantify the results for each scenario.

Oriented by a distinct objective, Cassettari et al. (2013) proposed a DES to evaluate costs impact and occupancy resources rates in CSM hospital integration. In their case, an Italian medium-sized local hospital conduced the same sterilization process in two different departments with no resource sharing where the main line located in the Central Sterilization Department (CSD), and a smaller line located in the Central Surgery Structure (CSS) for emergency cases.

Moreover, Cassettari's research considers that part of sterilization process is deterministic (cycle times of washing machines and autoclaves) and that there was no distinction between material characteristics (volume, for instance). Therefore, in this research its considered that all sterilization process has a stochastic nature, which is true even in machine process because we aim consider the total process time and it depends the materials processed in this cycle, the external temperature and another factors (AORN, 2012).

## BACKGROUND

### Discrete Event Simulation

Shannon (1975) defines simulation as "process of elaborating a model of a real (or hypothetical) system and conducting experiments with the purpose of understanding the behavior of a system or evaluating its operation."

Discrete Event Simulation (DES) consists in a model that, according to Law & Kelton (2000), can be defined as "a set of assumptions about how a system works, to try to gain some understanding of how the system behaves".

According to Sakurada & Miyake (2009) a system may be represented from different approaches (Figure 1). According to the same authors, Discrete Event Simulation covers the study of simulation models whose variables change state instantaneously at specific points of time, in contrast to what happens with continuous models, whose variables may change the state continuously along the time.

Despite the high efforts for data collection and development and validation time, a simulation model, according to Chwif (1999), has as advantages:

1. **Reduced Risk of Accidents:** In case there is any problem with the control Logic, it will be detected in the simulation, avoiding problems or material damage;
2. **Reduction of Problems That May Occur in the Field:** Since before testing the control logic in field, it is tested in the "laboratory", the possibility of still encountering logic errors is minimized;
3. **Extensive Tests:** With the simulation, a wide range of situations can be generated to test the control logic, making it "stronger", since it may work either for common situations as for uncommon ones, which may occur;

*Figure 1. Approaches for representation of productive systems*
*Source: Sakurada & Miyake (2009).*

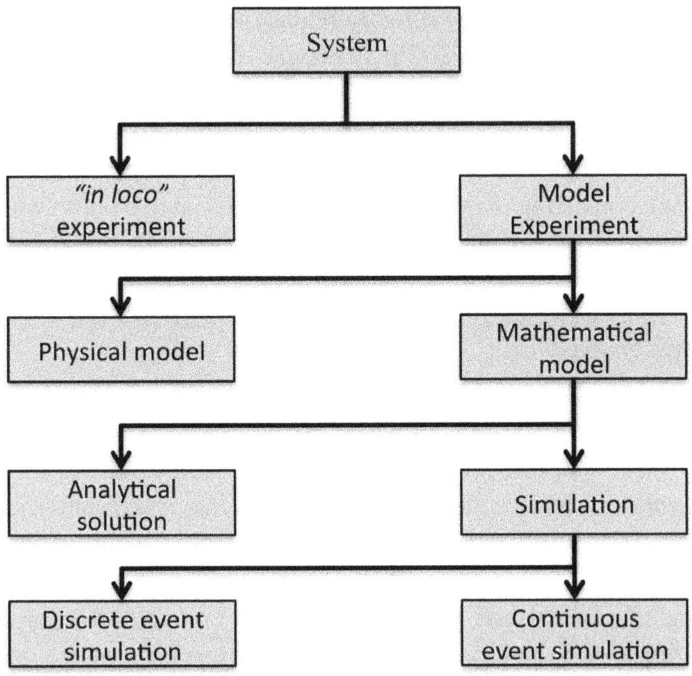

4. **The Cost of Each Simulation Round/Set-Ups, in the Simulation Model Is Despicable:** Once the simulation model was built and customized, the time cost of assembling new configurations is minimal (just some simulation in the keyboard). In practice to test logic in a new system configuration may generate costs since there is all the time real time and costs set-ups.

## CSM Operations Management

From the internal supply chain point of view (SPIEGEL et al, 2014), the CSM is one of the links that provide material for the other components of the hospital. CSM provides sterilized material for infirmary, operating rooms, transplant units and outpatient clinics (BASU et al, 2015), in other words, it is the area responsible for providing the other hospital sectors with the necessary inputs, instruments and clinical materials in compliance for patient safety. This way, if CSM has a delay in delivering the material, or a low speed of delivering or even deliver non conformity material that may be reprocessed, etc., it impacts straight in the hospital performance.

Concerning to operations, the sterilization activity performed at the CSM is essential for the proper working of medical care and testing. However, in the perspective of costs, it becomes an expensive activity, since to sterilize 0.5 $m^3$ of medical devices can cost more than 250 euros (DI MASCOLO & GOUIN, 2013). Reymondon, Pellet & Marcona (2008) relate high quality in the hospital service to hygienic conditions, which are directly related to the used surgical and hospital instruments. Di Mascolo et al.(2006) highlight that the sterilizations service is essential for the fights against the infections.

The processing of products for health (PPH) implies in a set of actions, which starts within the receipt of this contaminated product in the Center of Sterilized Material (CSM), consisting of: the receipt of the material to be processed, cleaning, drying, visual inspection, evaluation of the integrity and functionality, disinfection and/or sterilization, preparation (assembling and packing), storage, distribution and transport. It is produced, thereby, sterile article ready for safe use (PEREIRA, 2010; BRAZIL, 2012).

Each one of these steps of the PPH, with its relevance and complexity, demands a physical area in the CSM that provides conditions for the better development of the whole work. Besides, it requires equipment with advanced technology and professionals with appropriate training and skills to perform such activities (BRAZIL, 2002; BRAZIL, 2012; PEZZI & LEITE, 2010).

Normative, in the context of sterilization, have the objective of standardizing the processing of clinical and surgical materials. Leite (2008) emphasizes that the standardization of standards and technical routines is essential to validate the material processing and infection control. The CSM normative sets are composed of ANVISA's technical note, orders and resolutions RDC). There are the CSM International certification sets, which provide international certification, such as the IAHCSMM (International Association of Healthcare Central Services and Material Management) and the CBSPD (Certification Board for Sterile Processing and Distribution). There is the CSM accreditation, offered by Joint Commission on Accreditation of Healthcare Organizations. However, in none of these elements that support the definition of primary elements such as the dimensioning and allocation of essential resources for the management of the operations of a CSM (SPIEGEL et al.,2016).

# THE MODEL

## Context of the Studied CSM

The Pedro Ernesto University Hospital (HUPE) was inaugurated in 1950 with activities that privileged exclusively the academic questions of teaching and research, with the follow up and studies of clinical rarities and end-stage diseases and that in 1975 also became a general care hospital. The (HUPE) provides free public service and it's inserted in the Brazilian Health Unique System, the SUS. It currently receives patients of medium and high complexity from all the ECUs (Emergency Care Units), state hospitals and from other cities of the state. Today, one of the largest health care teaching complexes, with 516 beds, and a reference in a series of high complexity specialties and an important national nucleus of professionals training, in the medical area, linked to the University of the State of Rio de Janeiro (UERJ) in Brazil.

The hospital is managed by the Government of the State of Rio de Janeiro and faces a serious financial crisis, which places increasing challenges in the identification of alternatives for its operations management, exploring the potentialities with the already installed resources. With regard to CSM, there is an expectation that it needs additional human resources to meet the current demand. During the research, we evaluated this increment and showed that the critical resource is not the human one, but the autoclave. We went forward in the research and presented an alternative that does not prescind of investment and increase the installed capacity through a re-ordering of the material receiving schedule.

The HUPE's CSM operates as the value chain seen in Figure 2.

## Assumptions

Surgical entities are divided into 5 categories, which were properly validated by the CSM's professionals and are presented in Table 1. As parameters of the models,

*Figure 2. Structure of sterilization process*

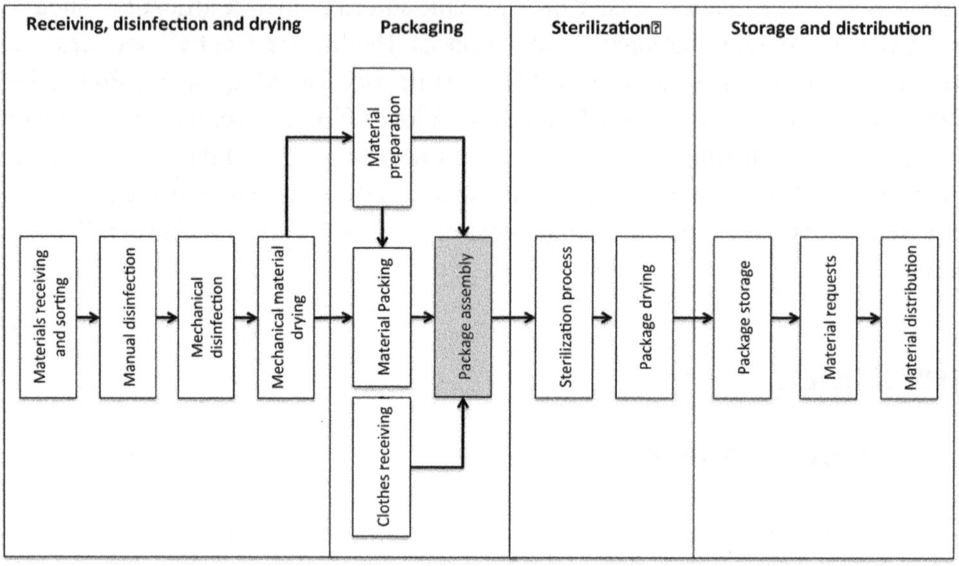


*Table 1. Surgical entities data used in the model*

| Category (Surgical Entities) - e | Number of Parts | Volume |
|---|---|---|
| Big box with many items | More than 50 parts | 0.5 baskets |
| Big box with few items | More than 50 parts | 0.5 baskets |
| Small box with many items | From 1 to 50 parts | 0.17 baskets |
| Small box with few items | From 1 to 50 parts | 0.17 baskets |
| Video components | Not applicable | 0.17 baskets |

the number of parts was chosen, because the higher this value the more time it will be necessary to complete the activities in the manual stations, which ha as a target the washing and disinfection (made piece by piece) and volume (necessary to accommodate the entities in the autoclave). Weight was disregarded since there is no limitations regarding to this characteristic.

The receiving of the surgical entities happens in the same time window where the surgery ends (e.g.: surgery finished at 9.35 has its material delivered between 9.00 and 10.00)

The non-surgical entities are described and their volume parameter is presented in the Table 2.

*Table 2. Non surgical entities data used in the model*

| Category (Non Surgical Entities) - e | Volume |
|---|---|
| Clothing and surgical field | 0.25 baskets |
| Clinical material | 0.25 baskets |

Reception times and any non-surgical entity is previously stablished and lasts 1 hour.

The machines present in the process have their capacities specified in Table 3.

From these descriptions it would be possible to move forward in optimization models to create theses batches. However, in practice, it's unlikely that health professionals will use such models in real time, for several reasons, such as: (i) Unavailability of computers in all posts; (ii) Incompatible professional training; (iii) Time restraint, etc.

Thus, in this work, choosing of the items that will compose the batches to be processed on each machine, is indeed, randomly, subject to the following restrictions:

$$\sum_{e} X\left(e,m\right) * Volume\left(e\right) \leq Capacity\left(m\right), \forall m \tag{1}$$

*Table 3. Equipment capacity data*

| Machine - m | Total Capacity | Minimum Occupancy |
|---|---|---|
| Autoclave 01 | 12 baskets | 10 baskets |
| Autoclave 02 | 12 baskets | 10 baskets |
| Autoclave 03 | 8 baskets | 7 baskets |
| Drying Machine | 5 baskets | 4 baskets |
| Ultrasonic wash | 8 baskets | 7 baskets |

*Table 4. Equipment capacity constrains*

| Set | Parameter | Variable |
|---|---|---|
| e – Surgical and non-surgical entities<br>m – machine | Capacity (m)<br>Minimum occupancy (m)<br>Volume (e) | X (e,m)– Number of entities and in the machine m |

$$\sum_{e} X(e,m) * Volume(e) \geq Minimumoccupancy(m), \forall m \quad (2)$$

In the inequality 1, it's assured that the amount of the items will not exceed the maximum volume of the machines, whereas in inequality 2, it's assures that for any batch composition, the entities will only be placed in the machines, when the volume of the material to be processed is greater than the minimum occupancy. The second inequality had as parameters information provided by the area professionals, who affirm, due to cost reasons, they work in these operating ranges (between 80% and 90% of the total capacity).

## Data

The data was obtained from the records in the CSM logbook. In it, it was observed that the non-surgical entities have distinct treatment regarding to the arrival times.

The distribution of surgical and non-surgical entities along the time ranges is presented in the Table 5.

Processing times per machine and type of entity were accrued by a team that measured these parameters, through in loco observations, for two months. These measurements synthesis is presented in Table 6 and the processing times by entity in each cleaning and sterilization stage were validated by CSM professionals. It was adopted the uniform distribution for machines, due to the processing time depend on the density of the load that forms the machine batches.

# SOLUTIONS AND RECOMMENDATIONS

According to the coordinator nurse of the CSM, the high times of crossing are related to the low number of professionals dedicated to the area, both in the shift of 7 to 19 o'clock and in the 19 to 7 o'clock. Currently, she has 5 professionals that could be picked up in two different areas. The first area comprises cleaning and packaging the used material to make it suitable for reprocessing. The second one comprises the removal of the sterilized material from the autoclave and its storage, for further distribution at the request of the different care services or surgical center. It's possible that at the same day, a professional located at the second area to be relocated to the first, however, the reverse is not allowed because of issues related to the contamination risk of the sterile material.

Thus, currently, the CSM coordinator nurse locates 4 professional in the area intended for cleaning of materials to be reprocessed ("dirty area") and 1 professional

*Table 5. Current material arrival rules*

| Time Slot | Large Box Few Items | Big Box Many Items | Small Box Few Items | Small Box Many Items | Clinical Material | Video Components | Clothing | Total |
|---|---|---|---|---|---|---|---|---|
| 0 to 1 | 0 | 0 | 0 | 0 | 0 | 0 | 0 | 0 |
| 1 to 2 | 0 | 0 | 0 | 0 | 0 | 0 | 0 | 0 |
| 2 to 3 | 0 | 0 | 0 | 0 | 0 | 0 | 0 | 0 |
| 3 to 4 | 0 | 0 | 0 | 0 | 0 | 0 | 0 | 0 |
| 4 to 5 | 0 | 0 | 0 | 0 | 0 | 0 | 0 | 0 |
| 5 to 6 | 0 | 0 | 0 | 0 | 0 | 0 | 0 | 0 |
| 6 to 7 | 0 | 0 | 0 | 0 | 0 | 0 | 0 | 0 |
| 7 to 8 | 0 | 0 | 0 | 0 | 0 | 0 | 0 | 0 |
| 8 to 9 | 0 | 1 | 0 | 1 | 0 | 0 | 0 | 2 |
| 9 to 10 | 0 | 2 | 0 | 2 | 0 | 0 | 1 | 5 |
| 10 to 11 | 0 | 4 | 0 | 3 | 25 | 0 | 3 | 35 |
| 11 to 12 | 2 | 3 | 2 | 2 | 0 | 0 | 5 | 14 |
| 12 to 13 | 1 | 3 | 1 | 2 | 0 | 7 | 5 | 19 |
| 13 to 14 | 1 | 3 | 1 | 2 | 0 | 0 | 4 | 11 |
| 14 to 15 | 1 | 3 | 1 | 2 | 0 | 0 | 4 | 11 |
| 15 to 16 | 0 | 3 | 0 | 2 | 25 | 0 | 4 | 34 |
| 16 to 17 | 0 | 3 | 0 | 2 | 0 | 14 | 4 | 23 |
| 17 to 18 | 0 | 3 | 0 | 2 | 0 | 0 | 4 | 9 |
| 18 to 19 | 0 | 3 | 0 | 0 | 0 | 0 | 3 | 6 |
| 19 to 20 | 0 | 2 | 0 | 0 | 0 | 0 | 2 | 4 |
| 20 to 21 | 0 | 1 | 0 | 0 | 0 | 0 | 1 | 2 |
| 21 to 22 | 0 | 0 | 0 | 0 | 25 | 0 | 0 | 25 |
| 22 to 23 | 0 | 0 | 0 | 0 | 0 | 0 | 0 | 0 |
| 23 to 0 | 0 | 0 | 0 | 0 | 0 | 0 | 0 | 0 |
| Total | 5 | 34 | 5 | 20 | 75 | 21 | 40 | 200 |

in the area intended to the storage and distribution of the sterile material ("clean area"). In addition, it was requested by the hospital management that the surgical entities were prioritized in relation to the non-surgical ones in order for these not to "steal the capacity" of CSM during the working hours of the operating room for elective procedures (delays of the surgery because the non-surgical entities were routinely processed in the autoclave while the surgical entities necessary for the

*Figure 3. Demand by hour in current materials arrival rules*

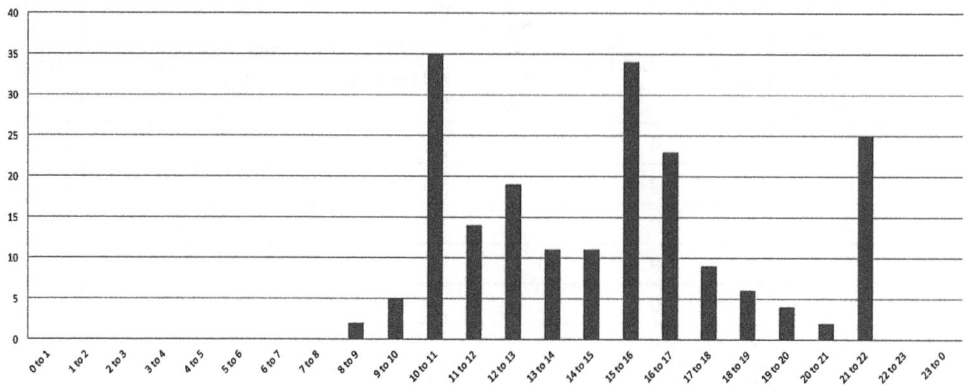


surgical procedure of an elective patient could just get in after the end of the batch in process).

Knowing the rules that govern the process, the number of people and with validation of the data presented in the previous section, the following question to be answered by discrete event simulation is simulated: "increase the human resources will impact (reduce) in a significant way the crossing times?"

Table 7 summarizes relevant indicators for the posed question. The simulation had a day horizon and were done 5.000 replications for each scenario. In the columns are presented: (i) the lower average value of all replications. (ii) the average value considering all the replications and (iii) the higher average values of all the replications.

From this chart on, one tried to increase the number of people in the dirty area (where the queue is longer) from 4 to 7 people in order to check if the crossing times could be reduced.

Comparing both scenarios, it's verified that the waiting and crossing times don't reduce significantly while the use of human resources in the dirty area falls by more than 42%.

As the only significant impact was the use of human resources in the dirty area, it's possible to state that, in the actual configuration, human resource is not the factor that limits production. Besides, in both scenarios it is possible to verify different levels of machines use and that the wait for processing in the autoclaves significantly impacts the crossing time of the entities, may just this wait vary between 34,1% in the lower average value replication to 52,52% in the higher average value replication, meaning, in the worst scenario, the wait for processing in the autoclave is responsible for more than half of the crossing time.

*Table 6. Data used in the model*

| Stage | Entities | Processing Time (in Minutes) | | | |
|---|---|---|---|---|---|
| | | Minimum Value | Medium | Maximum Value | Probability Distribution |
| Receiving | All | 2 | 2.5 | 3 | Triangular |
| Manual washing | Large box few items | 3 | 9.13 | 33 | Triangular |
| | Large box many items | 13 | 18.14 | 32 | Triangular |
| | Small box few items | 2 | 12 | 23 | Triangular |
| | Small box many items | 8 | 15 | 28 | Triangular |
| | Video Components | 3 | 8 | 20 | Triangular |
| Ultrasonic washing | All the entities, except for clothing. | 10 | Not applicable | 12 | Uniform |
| Preparation for drying | Large box few items | 1 | 1.79 | 4 | Triangular |
| | Large box many items | 1 | 1.75 | 3 | Triangular |
| | Small box few items | 1 | 1.4 | 3 | Triangular |
| | Small box many items | 1 | 1.6 | 5 | Triangular |
| | Video Components | 1 | 1.3 | 3 | Triangular |
| Dryer | All the entities, except for clothing. | 10 | Not applicable | 20 | Uniform |
| Preparation for packaging | Large box few items | 3 | 5 | 7 | Triangular |
| | Large box many items | 1 | 3.03 | 11 | Triangular |
| | Small box few items | 1.38 | 1.55 | 1.72 | Triangular |
| | Small box many items | 2.5 | 3 | 3.43 | Triangular |
| | Video Components | 20 | 30 | 40 | Triangular |
| | Clothing | 0.5 | 1.43 | 5 | Triangular |
| Sterilization | All the entities | 120 | Not applicable | 180 | Uniform |
| Storage and distribution | All the entities | 11 | 12 | 17 | Triangular |

It's been proved that increase human resources does not solve the problem and since it has been discovered the bottleneck of this production system for the current case mix, one may state that the acquisition of a new autoclave would reduce the average crossing times of the entities, however, there are some practical obstacles, such as: (i) Unavailability of physical space in the current configuration (it would be necessary to redefine the layout); (ii) It needs interventions that would temporarily interrupt the supply of the surgical center, infirmaries, outpatient clinics, among others; (iii) It is a resource of high value.

*Table 7. Current materials arrival rules results*

| Simulation Results | | Areas | Dirty | Clean |
|---|---|---|---|---|
| | | Resources | 4 | 1 |
| Evaluated Parameters | Entities | Scenario 1 | | |
| | | Minimum | Medium | Maximum |
| Processing time (VA time) in hours | Big box many items | 3,42 | 3,68 | 3,87 |
| | Big box few items | 3,09 | 3,55 | 4,10 |
| | Small box many items | 2,93 | 3,53 | 4,04 |
| | Small box few items | 3,16 | 3,47 | 3,70 |
| | video component | 3,59 | 3,92 | 4,20 |
| | Clinical material | 2,66 | 2,96 | 3,17 |
| | clothing | 2,64 | 2,89 | 3,06 |
| Wait time in hours | Big box many items | 2,27 | 3,99 | 5,70 |
| | Big box few items | 0,68 | 2,36 | 4,16 |
| | Small box many items | 0,68 | 1,85 | 4,14 |
| | Small box few items | 1,41 | 2,78 | 4,41 |
| | video component | 1,57 | 4,20 | 6,82 |
| | Clinical material | 1,20 | 3,11 | 4,82 |
| | clothing | 1,65 | 2,76 | 4,83 |
| Crossing time (total time) in hours | Big box many items | 5,97 | 7,68 | 9,40 |
| | Big box few items | 3,98 | 5,91 | 7,60 |
| | Small box many items | 3,98 | 5,38 | 7,99 |
| | Small box few items | 4,79 | 6,25 | 8,10 |
| | video component | 5,75 | 8,12 | 10,71 |
| | Clinical material | 4,08 | 6,07 | 7,81 |
| | clothing | 4,46 | 5,64 | 7,74 |
| Use by station (Scheduled Utilization) | Receiving | 0,12 | 0,17 | 0,22 |
| | Manual Washing | 0,17 | 0,23 | 0,32 |
| | Preparation for drying | 0,02 | 0,03 | 0,04 |
| | Preparation for packaging | 0,16 | 0,27 | 0,39 |
| | Storage and distribution | 0,16 | 0,20 | 0,24 |
| Use of machines (Number busy) | Autoclave 1 | 0,60 | 0,77 | 0,92 |
| | Autoclave 2 | 0,60 | 0,76 | 0,93 |
| | Autoclave 3 | 0,61 | 0,74 | 0,89 |
| | Ultrasonic washing | 0,10 | 0,13 | 0,18 |
| | Drying Machine | 0,13 | 0,18 | 0,23 |
| Waiting time in queue in hours | Wait for autoclave availability | 1,63 | 2,66 | 4,45 |

*Table 8. Adding professionals in dirty area following the current materials arrival rules results*

| Simulation Results | | Areas | Dirty | Clean |
|---|---|---|---|---|
| | | Resources | 7 | 1 |
| Evaluated Parameters | Entities | Scenario 2 | | |
| | | Minimum | Medium | Maximum |
| Processing time (VA time) in hours | Big box many items | 3,43 | 3,67 | 3,99 |
| | Big box few items | 3,20 | 3,59 | 4,00 |
| | Small box many items | 3,20 | 3,52 | 3,98 |
| | Small box few items | 3,20 | 3,47 | 3,72 |
| | video component | 3,57 | 3,92 | 4,25 |
| | Clinical material | 2,80 | 2,97 | 3,24 |
| | clothing | 2,63 | 2,89 | 3,06 |
| Wait time in hours | Big box many items | 2,15 | 4,02 | 6,12 |
| | Big box few items | 1,00 | 2,31 | 4,78 |
| | Small box many items | 1,00 | 1,70 | 3,79 |
| | Small box few items | 1,48 | 2,80 | 4,71 |
| | video component | 2,03 | 4,00 | 6,39 |
| | Clinical material | 1,44 | 3,06 | 4,84 |
| | clothing | 1,67 | 2,79 | 4,73 |
| Crossing time (total time) in hours | Big box many items | 5,73 | 7,69 | 9,91 |
| | Big box few items | 4,50 | 5,90 | 8,43 |
| | Small box many items | 4,50 | 5,22 | 7,29 |
| | Small box few items | 4,80 | 6,27 | 8,24 |
| | video component | 5,60 | 7,92 | 10,09 |
| | Clinical material | 4,37 | 6,03 | 7,78 |
| | clothing | 4,46 | 5,68 | 7,63 |
| Use by station (Scheduled Utilization) | Receiving | 0,07 | 0,10 | 0,12 |
| | Manual Washing | 0,10 | 0,13 | 0,17 |
| | Preparation for drying | 0,01 | 0,02 | 0,02 |
| | Preparation for packaging | 0,09 | 0,15 | 0,22 |
| | Storage and distribution | 0,15 | 0,20 | 0,24 |
| Use of machines (Number busy) | Autoclave 1 | 0,64 | 0,76 | 0,92 |
| | Autoclave 2 | 0,62 | 0,76 | 0,92 |
| | Autoclave 3 | 0,58 | 0,75 | 0,94 |
| | Ultrasonic washing | 0,10 | 0,13 | 0,16 |
| | Drying Machine | 0,13 | 0,18 | 0,23 |
| Waiting time in queue in hours | Wait for autoclave availability | 1,37 | 2,69 | 4,35 |

*Table 9. Comparing scenarios results the current materials arrival rules results*

| Simulation Results | | Areas | Dirty | Clean | Areas | Dirty | Clean |
|---|---|---|---|---|---|---|---|
| | | Resources | 4 | 1 | Resources | 7 | 1 |
| Evaluated Parameters | Entities | Scenario 1 | | | Scenario 2 | | |
| | | Minimum | Medium | Maximum | Minimum | Medium | Maximum |
| Processing time (VA time) in hours | Big box many items | 3,42 | 3,68 | 3,87 | 3,43 | 3,67 | 3,99 |
| | Big box few items | 3,09 | 3,55 | 4,10 | 3,20 | 3,59 | 4,00 |
| | Small box many items | 2,93 | 3,53 | 4,04 | 3,20 | 3,52 | 3,98 |
| | Small box few items | 3,16 | 3,47 | 3,70 | 3,20 | 3,47 | 3,72 |
| | video component | 3,59 | 3,92 | 4,20 | 3,57 | 3,92 | 4,25 |
| | Clinical material | 2,66 | 2,96 | 3,17 | 2,80 | 2,97 | 3,24 |
| | clothing | 2,64 | 2,89 | 3,06 | 2,63 | 2,89 | 3,06 |
| Wait time in hours | Big box many items | 2,27 | 3,99 | 5,70 | 2,15 | 4,02 | 6,12 |
| | Big box few items | 0,68 | 2,36 | 4,16 | 1,00 | 2,31 | 4,78 |
| | Small box many items | 0,68 | 1,85 | 4,14 | 1,00 | 1,70 | 3,79 |
| | Small box few items | 1,41 | 2,78 | 4,41 | 1,48 | 2,80 | 4,71 |
| | video component | 1,57 | 4,20 | 6,82 | 2,03 | 4,00 | 6,39 |
| | Clinical material | 1,20 | 3,11 | 4,82 | 1,44 | 3,06 | 4,84 |
| | clothing | 1,65 | 2,76 | 4,83 | 1,67 | 2,79 | 4,73 |

*continued on following page*

*Table 9. Continued*

| Simulation Results | | Areas | Dirty | Clean | Areas | Dirty | Clean |
|---|---|---|---|---|---|---|---|
| | | Resources | 4 | 1 | Resources | 7 | 1 |
| Evaluated Parameters | Entities | Scenario 1 | | | Scenario 2 | | |
| | | Minimum | Medium | Maximum | Minimum | Medium | Maximum |
| Crossing time (total time) in hours | G box many items | 5,97 | 7,68 | 9,40 | 5,73 | 7,69 | 9,91 |
| | G box few items | 3,98 | 5,91 | 7,60 | 4,50 | 5,90 | 8,43 |
| | p box many items | 3,98 | 5,38 | 7,99 | 4,50 | 5,22 | 7,29 |
| | P box few items | 4,79 | 6,25 | 8,10 | 4,80 | 6,27 | 8,24 |
| | video component | 5,75 | 8,12 | 10,71 | 5,60 | 7,92 | 10,09 |
| | Clinical material | 4,08 | 6,07 | 7,81 | 4,37 | 6,03 | 7,78 |
| | clothing | 4,46 | 5,64 | 7,74 | 4,46 | 5,68 | 7,63 |
| Use by station (Scheduled Utilization) | Receiving | 0,12 | 0,17 | 0,22 | 0,07 | 0,10 | 0,12 |
| | Manual Washing | 0,17 | 0,23 | 0,32 | 0,10 | 0,13 | 0,17 |
| | Preparation for drying | 0,02 | 0,03 | 0,04 | 0,01 | 0,02 | 0,02 |
| | Preparation for packaging | 0,16 | 0,27 | 0,39 | 0,09 | 0,15 | 0,22 |
| | Storage and distribution | 0,16 | 0,20 | 0,24 | 0,15 | 0,20 | 0,24 |
| Use of machines (Number busy) | Autoclave 1 | 0,60 | 0,77 | 0,92 | 0,64 | 0,76 | 0,92 |
| | Autoclave 2 | 0,60 | 0,76 | 0,93 | 0,62 | 0,76 | 0,92 |
| | Autoclave 3 | 0,61 | 0,74 | 0,89 | 0,58 | 0,75 | 0,94 |
| | Ultrasonic washing | 0,10 | 0,13 | 0,18 | 0,10 | 0,13 | 0,16 |
| | Drying Machine | 0,13 | 0,18 | 0,23 | 0,13 | 0,18 | 0,23 |
| Waiting time in queue in hours | Wait for autoclave availability | 1,63 | 2,66 | 4,45 | 1,37 | 2,69 | 4,35 |

From the results of the two scenarios it was possible to answer the question proposed at the beginning of this section and extrapolate it to know which resource should be increased to significantly impact (reduce) the crossing times. However, facing the issues above mentioned, one may reformulate the question in this way: "is investing in the increase in resource capacity which limits production the only way of reducing the crossing time?"

In the Table 9, it's seen that the use of autoclaves is between 60% and 94% with an average value of approximately 75% for the current demand configuration and a new propose to reduce the average crossing times may be to specify new receiving times for the non-surgeries entities aiming to increase the use of the critical resource and, consequently, reduce the average crossing times. For this proposal, there are many possible combinations, but it is not the objective of this work to try them exhaustively. Thus, it's proposed the following configuration (Table 10).

Changing the materials arrival rules for the proposed configuration above, one reached the results presented in Table 11.

Although there is no guarantee of optimality, the adoption of the receiving time of non-surgical entities, proposed in this scenario, has as benefits: (i) reducing the crossing time of all entities from the decrease in the average waiting time for autoclave processing; (ii) better use of the critical resource.

## FUTURE RESEARCH DIRECTIONS

As future resources it's recommended the continuity of the research, advancing in optimization models, which improves the definition of the best time to receive non-surgical entities, aiming the maximization of the CSM resources use.

*Figure 4. Demand by hour in materials arrival rules proposed*

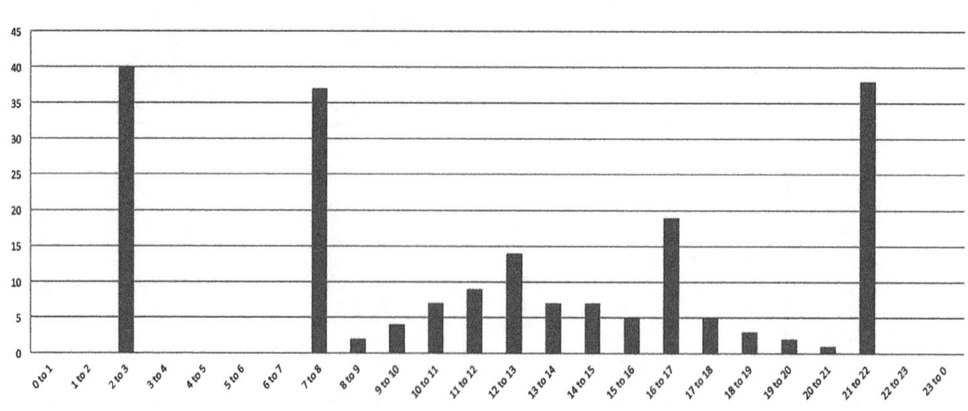

*Table 10. Materials arrival rules proposed*

| Hour Range | Big Box Few Items | Big Box Many Items | Small Box Few Items | Small Box Many Items | Clinical Material | Video Components | Clothing | Total |
|---|---|---|---|---|---|---|---|---|
| 0 to 1 | 0 | 0 | 0 | 0 | 0 | 0 | 0 | 0 |
| 1 to 2 | 0 | 0 | 0 | 0 | 0 | 0 | 0 | 0 |
| 2 to 3 | 0 | 0 | 0 | 0 | 0 | 0 | 40 | 40 |
| 3 to 4 | 0 | 0 | 0 | 0 | 0 | 0 | 0 | 0 |
| 4 to 5 | 0 | 0 | 0 | 0 | 0 | 0 | 0 | 0 |
| 5 to 6 | 0 | 0 | 0 | 0 | 0 | 0 | 0 | 0 |
| 6 to 7 | 0 | 0 | 0 | 0 | 0 | 0 | 0 | 0 |
| 7 to 8 | 0 | 0 | 0 | 0 | 37 | 0 | 0 | 37 |
| 8 to 9 | 0 | 1 | 0 | 1 | 0 | 0 | 0 | 2 |
| 9 to 10 | 0 | 2 | 0 | 2 | 0 | 0 | 0 | 4 |
| 10 to 11 | 0 | 4 | 0 | 3 | 0 | 0 | 0 | 7 |
| 11 to 12 | 2 | 3 | 2 | 2 | 0 | 0 | 0 | 9 |
| 12 to 13 | 1 | 3 | 1 | 2 | 0 | 7 | 0 | 14 |
| 13 to 14 | 1 | 3 | 1 | 2 | 0 | 0 | 0 | 7 |
| 14 to 15 | 1 | 3 | 1 | 2 | 0 | 0 | 0 | 7 |
| 15 to 16 | 0 | 3 | 0 | 2 | 0 | 0 | 0 | 5 |
| 16 to 17 | 0 | 3 | 0 | 2 | 0 | 14 | 0 | 19 |
| 17 to 18 | 0 | 3 | 0 | 2 | 0 | 0 | 0 | 5 |
| 18 to 19 | 0 | 3 | 0 | 0 | 0 | 0 | 0 | 3 |
| 19 to 20 | 0 | 2 | 0 | 0 | 0 | 0 | 0 | 2 |
| 20 to 21 | 0 | 1 | 0 | 0 | 0 | 0 | 0 | 1 |
| 21 to 22 | 0 | 0 | 0 | 0 | 38 | 0 | 0 | 38 |
| 22 to 23 | 0 | 0 | 0 | 0 | 0 | 0 | 0 | 0 |
| 23 to 0 | 0 | 0 | 0 | 0 | 0 | 0 | 0 | 0 |
| total | 5 | 34 | 5 | 20 | 75 | 21 | 40 | 200 |

Extending the objective of improving resource utilization and from a cost perspective (avoid duplication of human resources and equipment beyond physical space), it is proposed to create simulation models where different hospitals can share the same CSM. In this type of problem should be considered transport time uncertainties where it is suggested that the delivery sequence of the sterile materials be made by optimization models (this problem is typically called as vehicle routing

*Table 11. Proposed materials arrival rules results*

| Simulation Results | | Areas | Dirty | Clean |
|---|---|---|---|---|
| | | Resources | 4 | 1 |
| Evaluated Parameters | Entities | Scenario 3 | | |
| | | Minimum | Medium | Maximum |
| Processing time (VA time) in hours | Big box many items | 3,35 | 3,66 | 3,92 |
| | Big box few items | 3,17 | 3,58 | 4,01 |
| | Small box many items | 3,10 | 3,59 | 4,09 |
| | Small box few items | 3,18 | 3,46 | 3,73 |
| | video component | 3,52 | 3,89 | 4,32 |
| | Clinical material | 2,64 | 2,96 | 3,19 |
| | clothing | 2,62 | 2,87 | 3,11 |
| Wait time in hours | Big box many items | 0,86 | 2,75 | 5,30 |
| | Big box few items | 0,25 | 2,04 | 4,55 |
| | Small box many items | 0,50 | 2,12 | 5,02 |
| | Small box few items | 0,65 | 1,97 | 4,69 |
| | video component | 0,57 | 1,75 | 3,91 |
| | Clinical material | 1,13 | 2,21 | 3,89 |
| | clothing | 0,59 | 1,40 | 2,57 |
| Crossing time (total time) in hours | Big box many items | 4,44 | 6,42 | 8,94 |
| | Big box few items | 4,14 | 5,62 | 7,95 |
| | Small box many items | 3,97 | 5,71 | 8,63 |
| | Small box few items | 4,21 | 5,43 | 8,25 |
| | video component | 4,17 | 5,63 | 7,86 |
| | Clinical material | 4,23 | 5,17 | 6,80 |
| | clothing | 3,34 | 4,27 | 5,56 |
| Use by station (Scheduled Utilization) | Receiving | 0,11 | 0,17 | 0,23 |
| | Manual Washing | 0,16 | 0,23 | 0,29 |
| | Preparation for drying | 0,02 | 0,03 | 0,03 |
| | Preparation for packaging | 0,20 | 0,33 | 0,49 |
| | Storage and distribution | 0,19 | 0,24 | 0,27 |
| Use of machines (Number busy) | Autoclave 1 | 0,73 | 0,90 | 0,99 |
| | Autoclave 2 | 0,73 | 0,89 | 1,00 |
| | Autoclave 3 | 0,65 | 0,86 | 1,00 |
| | Ultrasonic washing | 0,10 | 0,13 | 0,16 |
| | Drying Machine | 0,13 | 0,18 | 0,22 |
| Waiting time in queue in hours | Wait for autoclave availability | 0,49 | 1,39 | 3,01 |

*Table 12. Comparing scenarios results between current and proposed materials arrival rules results*

| Simulation Results | | Increasing of Professionals | | | Change in the Arrival Rules | | |
|---|---|---|---|---|---|---|---|
| | | Percentage Difference Between the Times of Scenarios 1 and 2 | | | Percentage Difference Between the Times of Scenarios 1 and 3 | | |
| **Evaluated Parameters** | **Entities** | **Minimum** | **Medium** | **Maximum** | **Minimum** | **Medium** | **Maximum** |
| Processing time (VA time) in hours | Big box many items | 0,20% | -0,49% | 3,31% | -2,11% | -0,55% | 1,27% |
| | Big box few items | 3,55% | 0,96% | -2,49% | 2,41% | 0,59% | -2,14% |
| | Small box many items | 9,49% | -0,32% | -1,41% | 5,90% | 1,46% | 1,18% |
| | Small box few items | 1,45% | 0,01% | 0,46% | 0,80% | -0,18% | 0,69% |
| | video component | -0,62% | -0,08% | 1,12% | -1,93% | -0,91% | 2,84% |
| | Clinical material | 5,28% | 0,07% | 2,06% | -0,49% | -0,04% | 0,33% |
| | clothing | -0,62% | 0,20% | -0,05% | -0,88% | -0,67% | 1,69% |
| Wait time in hours | Big box many items | -5,34% | 0,71% | 7,24% | -62,16% | -31,01% | -7,05% |
| | Big box few items | 48,52% | -2,03% | 14,84% | -63,53% | -13,43% | 9,33% |
| | Small box many items | 48,52% | -7,89% | -8,49% | -25,81% | 14,88% | 21,20% |
| | Small box few items | 5,24% | 0,86% | 6,90% | -53,80% | -29,17% | 6,34% |
| | video component | 29,37% | -4,81% | -6,20% | -63,41% | -58,43% | -42,71% |
| | Clinical material | 19,13% | -1,43% | 0,42% | -5,94% | -28,98% | -19,46% |
| | clothing | 1,06% | 1,22% | -2,09% | -64,30% | -49,25% | -46,70% |

*continued on following page*

*Table 12. Continued*

| Simulation Results | | Increasing of Professionals | | | Change in the Arrival Rules | | |
|---|---|---|---|---|---|---|---|
| | | Percentage Difference Between the Times of Scenarios 1 and 2 | | | Percentage Difference Between the Times of Scenarios 1 and 3 | | |
| **Evaluated Parameters** | **Entities** | **Minimum** | **Medium** | **Maximum** | **Minimum** | **Medium** | **Maximum** |
| Crossing time (total time) in hours | Big box many items | -4,05% | 0,14% | 5,46% | -25,60% | -16,39% | -4,91% |
| | Big box few items | 13,01% | -0,23% | 10,87% | 3,97% | -5,00% | 4,55% |
| | Small box many items | 13,01% | -2,92% | -8,83% | -0,31% | 6,07% | 7,93% |
| | Small box few items | 0,28% | 0,39% | 1,82% | -11,98% | -13,08% | 1,95% |
| | video component | -2,69% | -2,53% | -5,75% | -27,59% | -30,65% | -26,61% |
| | Clinical material | 7,11% | -0,70% | -0,35% | 3,65% | -14,85% | -12,86% |
| | clothing | -0,02% | 0,70% | -1,51% | -25,05% | -24,40% | -28,19% |
| Use by station (Scheduled Utilization) | Receiving | -41,86% | -42,59% | -43,58% | -5,98% | 0,41% | 5,06% |
| | Manual Washing | -41,80% | -42,67% | -48,29% | -4,62% | 0,56% | -9,09% |
| | Preparation for drying | -43,03% | -41,93% | -44,70% | -1,48% | 1,22% | -9,17% |
| | Preparation for packaging | -39,89% | -42,52% | -43,60% | 27,40% | 24,97% | 23,84% |
| | Storage and distribution | -6,87% | 0,20% | 0,55% | 19,57% | 18,52% | 12,63% |
| Use of machines (Number busy) | Autoclave 1 | 5,57% | -0,16% | 0,63% | 20,87% | 17,07% | 8,30% |
| | Autoclave 2 | 3,44% | -0,24% | -1,38% | 20,53% | 16,61% | 6,99% |
| | Autoclave 3 | -3,82% | 1,05% | 5,68% | 6,42% | 15,83% | 12,54% |
| | Ultrasonic washing | -1,50% | 1,08% | -6,55% | -4,70% | 0,93% | -9,45% |
| | Drying Machine | -1,44% | 0,79% | 2,53% | -0,15% | 0,96% | -2,14% |
| Waiting time in queue in hours | Wait for autoclave availability | -15,92% | 1,27% | -2,21% | -69,78% | -47,58% | -32,48% |

problem in literature). In order to support this proposal, it is necessary to determine the characteristics of the materials to be transported, what kind of packaging they need to have according to the current normative instruments, what type of vehicle is best suited to carry this transport and what are their capacities (in terms Volume and weight, for example). From these surveys it would be possible to assess the impact of delivery times and arrival rules and their respective fixed and variable costs.

## CONCLUSION

Given the relevance of CSM to hospital operation, this work sought to evaluate whether "will increasing human resources impact (reduce) significantly the crossing times?" Through simulation, one may see that the increase of the human resources does not impact in the crossing time, since the waiting time of the items to enter the autoclave, had in average 2,7 hours, whether in scenarios with 4 or 7 people allocated in the dirty area. Additionally, it was found that the critical resource would be stopped approximately 25% of the time.

From the results of the two scenarios, it was possible to answer the question initially proposed and extrapolate it to know which resource should be increased in order to impact (reduce) significantly, the crossing times. Faced with the infrastructural limitations and investment resources present in the reality of the studied hospital, the research continued, rephrasing the question to "is investing in the increasing of the resource capacity, which limits the production, the only way to reduce the crossing time?"

It was found that the adoption of reception policies different from the current ones, made it possible to make a better use of the critical resource, what reduces the system waiting time without any contracts, acquisitions, works or changes in the process.

The contributions of this research to the management of the health unit operations, brings, in the midst of the financial crisis experienced by it, the possibility of meeting more and better the internal demand, without any financial contribution. It's about showing the potentiality of bringing together professionals and academics, applying methods and management tools in the health sector.

## REFERENCES

AORN. (2012). Recommended Practices for Sterilization in the Perioperative Practice Setting. In Perioperative Standards and Recommended Practices. Denver, CO: AORN.

Araujo, C. A. S. (2005). *Fatores a serem gerenciados para o alcance da qualidade para os clientes internos: um estudo em um conjunto de hospitais brasileiros*. Rio de Janeiro: Tese de Doutorado em Administração. Universidade Federal do Rio de Janeiro.

Basu, D., Bhattacharya, S., Mahajan, A., Ramanan, V. R., & Chandy, M. (2015). Sterilization indicators in central sterile supply department: quality assurance and cost implications. *Infection Control & Hospital Epidemiology, 36*(4), 484-486.

Brasil. (2012). *Ministério da Saúde. Agência Nacional de Vigilância Sanitária (ANVISA), Resolução da Diretoria Colegiada (RDC) n. 15 de 15 março de 2012*. Diário Oficial da República Federativa do Brasil, Brasília.

BRAZIL. (2002). *Ministério da Saúde.Agência Nacional de Vigilância Sanitária (ANVISA), Resolução da Diretoria Colegiada (RDC), n. 50 de 21 de fevereiro de 2002*. Brasília: Diário Oficial da República Federativa do Brasil.

Capkun, V., Messner, M., & Rissbacher, C. (2012). Service specialization and operational performance in hospitals. *International Journal of Operations & Production Management, 32*(4), 468–495. doi:10.1108/01443571211223103

Cardoen, B., Demeulemeester, E., & Beliën, J. (2010). Operating room planning and scheduling: A classification scheme. *International Journal of Social Health Information Management, 1*(1), 71–83.

Cassettari, L., Mosca, M., Mosca, R., & Rolando, F. (2013). An healthcare process reengineering using discrete event simulation. *Proceedings of World Congress on Engineering and Computer Science*, 2, 1174-1179.

Chwif, L. (1999). Utilizando simulação de eventos discretos em projetos de sistemas automatizados de manufatura. São Paulo: Academic Press.

Demeulemeester, E., Beliën, J., Cardoen, B., & Samudra, M. (2013). Operating room planning and scheduling. In *Handbook of healthcare operations management* (pp. 121–152). Springer New York. doi:10.1007/978-1-4614-5885-2_5

Di Mascolo, M., & Gouin, A. (2013). A generic simulation model to assess the performance of sterilization services in health establishments. *Health Care Management Science, 16*(1), 45–61. doi:10.1007/s10729-012-9210-2 PMID:22886097

Di Mascolo, M., Gouin, A., & Cong, K. N. (2006). Organization of the production of sterile medical devices. *IFAC Proceedings, 39*(3), 35-40.

Guerriero, F., & Guido, R. (2011). Operational research in the management of the operating theatre: A survey. *Health Care Management Science*, *14*(1), 89–114. doi:10.1007/s10729-010-9143-6 PMID:21103939

Günal, M. M., & Pidd, M. (2010). Discrete event simulation for performance modelling in health care: A review of the literature. *Journal of Simulation*, *4*(1), 42–51. doi:10.1057/jos.2009.25

Hulshof, P. J., Kortbeek, N., Boucherie, R. J., Hans, E. W., & Bakker, P. J. (2012). Taxonomic classification of planning decisions in health care: A structured review of the state of the art in OR/MS. *Health Systems (Basingstoke, England)*, *1*(2), 129–175. doi:10.1057/hs.2012.18

Law, A. M., & Kelton, W. D. (2000). *Simulation Modeling and Analysis*. Singapore: McGraw-Hill.

Leite, F. B. (2008). *Central de Material Esterilizado Projeto de Reestruturação e Ampliação do Hospital Regional de Francisco Sá*. Centro Universitário Euroamericano–UNIEURO.

Malagón-Londoño, G., Morera, R. G., Laverde, G. P., & Paulo, A. F. D. (2003). *Administração hospitalar. 2*. Buenos Aires: Editorial Médica Panamericana.

Pereira, M. C. O. (2010). *Preparo, embalagem, transporte e armazenagem. In Limpeza, desinfecção e esterilização de artigos em serviços de saúde* (pp. 83–107). São Paulo: APECIH.

Perkins, J. J. (1983). *Principles and methods of sterilization in health sciences* (2nd ed.). Charles C. Thomas Pub.

Pezzi, M. C. S., & Leite, J. L. (2010). Investigação em central de material e esterilização utilizando a teoria fundamentada em dados. *Revista Brasileira de Enfermagem*, *63*(3), 391–396. doi:10.1590/S0034-71672010000300007 PMID:20658072

Pittet, D., Allegranzi, B., Storr, J., Nejad, S. B., Dziekan, G., Leotsakos, A., & Donaldson, L. (2008). Infection control as a major World Health Organization priority for developing countries. *The Journal of Hospital Infection*, *68*(4), 285–292. doi:10.1016/j.jhin.2007.12.013 PMID:18329137

Possari, J. F. (2010). *Assistência de enfermagem na recuperação anestésica*. São Paulo: Iátria.

Reymondon, F., Pellet, B., & Marcon, E. (2008). Optimization of hospital sterilization costs proposing new grouping choices of medical devices into packages. *International Journal of Production Economics*, *112*(1), 326–335. doi:10.1016/j.ijpe.2006.12.066

Sakurada, N., & Miyake, D. I. (2009). Aplicação de simuladores de eventos discretos no processo de modelagem de sistemas de operações de serviços. *Gestão & Produção*, *16*(1), 25–43. doi:10.1590/S0104-530X2009000100004

Seavey, R. N. (2007). Unloading, Storage, Distribuition, Transportation and Aseptic Presentation of sterile items. *Managing Infection Control, Denver*, *70*(9), 88–101.

Shannon, R. E. (1975). *Systems Simulation: The Art and Science*. Prentice-Hall.

Silva, S. B., Caulliraux, H. M., & Spiegel, T. (2015). Modelo de simulação para auxílio em planejamento e gestão de serviços de saúde. *S & G. Sistemas & Gestão*, *10*(4), 608–616. doi:10.20985/1980-5160.2015.v10n4.616

SOBECC. (2013). *Recuperação anestésica e Centro de Material e Esterilização: práticas recomendadas*. São Paulo: Sociedade Brasileira de Enfermeiros de Centro Cirúrgico.

Spiegel, T., & Assad, D. B. N. (2016). Operations Project and Management in Trauma Centers: The Case of Brazilian Units. In Handbook of Research on Healthcare Administration and Management. IGI Global.

Spiegel, T., & Cameira, R. F. (2016). The Processing Chain of Sterilized Material from the Health Units of Rio De Janeiro County: Analysis and Proposition of Scenarios. In Supply Chain Management: Applications for Manufacturing and Service Industries. New York: Nova Science Publishers.

Spiegel, T., Caulliraux, H. M., Treistman, F. E., Guimaraes, G. E., Vaccaro, G. L. R., Amorim, L. G., . . . Azzam, S. M. (2016). Projeto e Gestão de Operações em Saúde. In Tópicos Emergentes e Desafio Metodológicos em Engenharia de Produção: Casos, Experiências e Proposições. Rio de Janeiro: ABEPRO.

Spiegel, T., Vasconcelos, P. E. M., Porto, D. L., & Caulliraux, H. M. (2014). Supply Chain Integration Researches: An Overview of the Field. *International Journal of Supply Chain Management*, *3*, 12–20.

Sundbo, J. (1994). Modulization of service production and a thesis of convergence between service and manufacturing organizations. *Scandinavian Journal of Management*, *10*(3), 245-266.

Vähätalo, M., & Kallio, T. J. (2015). Organising health services through modularity. *International Journal of Operations & Production Management*, *35*(6), 925–945. doi:10.1108/IJOPM-12-2013-0523

Wilson, J. T. (1981). Implementation of computer simulation projects in health care. *The Journal of the Operational Research Society*, *32*(9), 825–832. doi:10.1057/jors.1981.161 PMID:10252405

## KEY TERMS AND DEFINITIONS

**Center of Sterilized Material:** Standalone unit responsible for supplying, preparing, distributing and managing the sterile medical devices.

**Discrete Event Simulation:** A model that set of assumptions about how a system works, to try to gain some understanding of how the system behaves.

**Healthcare Operations Management:** The decisions and actions that occur within the limits defined by the operating system design. These include activities such as the implementation of policies, procedures and strategies, contingent decision-making, process coordination, problem identification and resolution, response to uncertainty and unforeseen problems and rewarding people.

**Hospital Structure:** The structure that fulfill health care, administrative and formative functions to delivery quality health service to the population.

**Operations Management:** A multidisciplinary field that investigates the design, management and improvement of processes aimed at the development, production, distribution and delivery of products and services.

**Processing of Products for Health:** Consisting of: the receipt of the material to be processed, cleaning, drying, visual inspection, evaluation of the integrity and functionality, disinfection and/or sterilization, preparation (assembling and packing), storage, distribution and transport.

**Simulation:** Process of elaborating a model of a real (or hypothetical) system and conducting experiments with the purpose of understanding the behavior of a system or evaluating its operation.

# Chapter 6
# Capacity Decision in Emergency Hospital Operation Rooms:
## Sizing Using Simulation

**Ana Carolina Pereira de Vasconcelos Silva**
*Rio de Janeiro State University, Brazil*

**Daniel Bouzon Nagem Assad**
*Rio de Janeiro State University, Brazil*

**Thais Spiegel**
*Rio de Janeiro State University, Brazil*

## ABSTRACT

*The operations management is a multidisciplinary field that investigates, for instance, the design, management and processes improvement focused on the development, production, distribution and delivery of products and services, encompassing activities such as the implementation of policies, making quota decisions, identification and problem solving, response to uncertainty, among others. Regarding the resources dimensioning in hospitals, the Brazilian scenario is limited to legislative instruments that assume a prior and added sizing. This chapter uses a discrete event simulation tool to set the amount of operation rooms needed for patient care in an emergency department, so that emergency patients have guaranteed compliance, minimizing the cancellation of elective surgeries because of this type of demand. As a result, it was found that the minimum amount established by normative instruments was not appropriate to the specific requirements of the organization.*

DOI: 10.4018/978-1-5225-3226-2.ch006

## INTRODUCTION

In Brazil, as a constitutional principle, people have the right to health, which is a State duty, without limits (BRASIL, 1988). However, even considering the pertinence of this universal right, pragmatic differences and difficulties arise when the origins of the resources needed to support this right are questioned or not (TIEGHI, 2013). As defined by Hollnagel et al. (2013: 59), health care is “an open, extensive and widely effective system characterized by large numbers of people, emerging and adaptive behaviors over time.”

In this context, designing healthcare operations refers to a complex object, which includes, for example, making decisions about the size of the health unit and its location, which healthcare flows will be made available, what is the stock policy of hospital materials, which IT system to implement according to the regulatory apparatus, among others (SPIEGEL & CAMEIRA, 2016). The healthcare operations management refers to decisions and actions that occur within the limits defined by the operating system design. These include activities such as the implementation of policies, procedures and strategies, contingent decision-making, process coordination, problem identification and resolution, response to uncertainty and unforeseen problems, and rewarding people (SPIEGEL et al, 2016). Improving the system refers to experimentation and learning activities aimed to improve operational performance over time (GINO & PISANO, 2008: 6). In some particular cases, such as emergency hospital units, the system is subject to demand peaks triggered by external and therefore uncontrollable events, where each patient presents a unique set of needs (Smith et al. 2007). Thus, this system needs to be designed to deal with variability in a scenario where there is an incompatibility between the investments and their potential demand (HALL, 2013).

About the hospital units resources dimensioning, the Brazilian scenario is defined mainly by the Resolution of the Board of Directors (RDC) nº 50/2002, which, in the context of an emergency unit, for example, defines as minimum number of operating rooms intended for the emergency patients care (polytrauma, cardiac arrest, among others) only one room. The sizing imposed by normative instruments assumes a fixed and aggregate demand, without considering the variabilities and specificities of the unit in question (demand profile, characteristic of professionals working in the unit, availability of resources in real time, among others) (SPIEGEL & ASSAD, 2016). Although on a propositional bias, SOMASUS (2011) points out the need for both polytrauma and emergency rooms to be sized to serve at least two patients at the same time. To tackle this inconsistency, this article proposes to use a Discrete Event Simulation Tool to analyze the trade-offs between different emergency rooms availability policies, in order to subsidize the decision-making about the configuration of an operation theater. The choice of this tool is due to its adequacy for the studies in which variables change state instantaneously at specific points of time, in contrast to what happens with continuous models, whose variables may change the state continuously along the time.

## BACKGROUND

Operations management is a multidisciplinary field that investigates the design, management and improvement of processes aimed at the development, production, distribution and delivery of products and services (SPIEGEL, 2013). It involves a whole range of different decisions that will determine the overall purpose of the operation, structure and operational practices, where the organization's resources need to be managed in the way they are directed, how they are designed, how delivery is planned and controlled and how they are developed and improved (SLACK et al., 2009). Thus, operations management deals with decision-making related to production processes to ensure that the resulting products or services are produced according to the specifications, in the quantities and schedule required and with minimal cost (SHIN & SIEGEL, 1999).

As characterized in the previous section, health is a complex object. The high volume of patients, the multiple flows, inputs, equipment and procedures into the health system, the increasingly specialized medical knowledge, the rapid development of technology pushing the frontier of knowledge and creating pressures for constant updating of clinical engineering, the quick increase in care costs, all of these factors make the design and management of health organizations a challenge without parallel in other sectors of economic activity. The result is increasing queues, fragmented assistance, dissatisfied users and professionals, societal pressure for improvement, and ample gap for increasing productivity (HOPP & LOVEJOY, 2012). Pedroso & Malik (2011) highlight some relevant issues with regard to health management. The first is the extent of the health scope, ranging from a positive perspective (well-being) to a pathological perspective, so that the impact on life reinforces its social relevance. They also highlight its economic importance, since inefficiencies in health management mean considerable waste of financial resources. In addition, because its fragmented value chain, it can lead to local decisions that do not increase the value generated. In this way, the challenge is to maximize results and minimize costs, aligning the design and management of health operations with the complexity of their products and services, the stage of medical maturity knowledge, among other variables (SPIEGEL et al, 2016).

In this context, hospitals deal with essential assets, which are the health and life of patients. They fulfill health care, administrative and formative functions to deliver quality health service to the population and it is inserted, by the Brazilian standard, into a health system that is divided into public (SUS) and private, in which the same provider can fulfill functions for both subsystems. Jacques (2007) points out that when they work in an open system, they are affected by cyclical or seasonal demands and by the epidemiological characteristics of the patients, as well as operating with a fixed and indirect cost structure and living in a dynamic environment through the incessant discovery of new knowledge and technology. In a convergent way, for Vecina Neto & Malik (2011), hospital structures have peculiar characteristics, since market laws do not apply mechanically to the sector, because of human needs and non-market priorities, which are imposed independently of production costs, market value and prices practiced. In addition, they point out that competition is not a strong element in the environment of these organizations, since it is a chronically lacking resources segment for the population and, also, a lacking financial resources segment. They reinforce the question of demand variability and argue that each patient behaves in a different way, which hinders a rigid standardization of the work process and a rationalization of service provision. In addition, they also highlight the asymmetry of information, due to clients' difficulties in judging their treatments or needs in a environment where the characteristic of the simultaneous consumption to the production, marked in the service delivery, allow no time for previous control of the quality, nor stock to fit offering and demand. At least, they also raise the issue of the wide variety of professionals at different educational and formative levels and with distinct corporate interests and resistance of physicians to management programs (SILVA et al, 2016).

In this hospital structure, there is the emergency department, with a peculiar characteristic: patients simply "appear". This department may be one of the hospital's entrance doors that is subject to acute peaks triggered by external, and therefore uncontrollable, events where each patient presents a unique set of needs (SMITH et al., 2007). Given this, the need to design it to deal with all these variabilities emerges in a scenario where there is an incompatibility of investments for emergency hospitals and their potential demand (HALL, 2013). Hopp & Lovejoy (2013) point out two distinct sources of variability in the emergency department, namely: variability in patient entrance and treatment time. The surgical center of a hospital is the hospital unit responsible for ensuring the care of surgical patients. Mantoani (2012) distinguishes surgeries as emergency, urgency or elective. Emergency surgeries are those that, due to the severity of the patient's clinical condition, require immediate surgical intervention. Urgency surgeries require immediate intervention, allowing them to wait a few hours, in which the patient is kept under evaluation and clinical and laboratorial observation. Elective surgeries include surgical procedures that, even if indicated to treat the patient's clinical condition, can be performed on a pre-scheduled date.

The challenges they face, in general, are long patient waiting lists and inefficiently resource use (NIU et al., 2013). For Marques & Captivo (2015), the direct relationship with other departments make the efficiency in its management a central part for the financial sustainability of the hospital.

## CAPACITY DECISION IN AN EMERGENCY HOSPITAL OPERATION THEATER

### Problem Delimitation

The unpredictable nature of emergency demand and its need for care in the short term creates a trade-off between resource allocation for non-elective (emergency or urgent) and elective patients. Elective surgeries can be scheduled in advance, so that hospitals with a high level of efficiency need to ensure high resource utilization rates, acceptable patient waiting times, and reduced turnover times (SILVA et al, 2016). However, for hospitals that intend to comply with both profiles, a contingency plan is required, so available capacity must be guaranteed for immediate care. The operation theater planning is complex mainly due to the different sources of variability. Van Riet & Demeulemeester (2015) point out that these variabilities can not be ignored, since they influence the trade-offs between hospital costs and efficiency, on the one hand, and quality of care and patients' preferences, on the other. Lamiri et al. (2008) proposes a stochastic mathematical model to treat both demands in order to minimize the relative costs for elective patient surgery.

According to Litvak (2010), a hospital system can be affected by natural and artificial variabilities. The natural variabilities are those whose parameters are not measurable, as the professional variability (asymmetry of knowledge, experience and skills), clinical variability (patients come with different diseases, different clinical status and different responses even when they are subjected to the same treatment) and entrance rate variability (patients do not reach a certain rate, but when they need medical assistance). Natural variabilities are therefore random nature variabilities and should be managed on a day-to-day basis. Artificial variabilities have measurable parameters, such as those associated with scheduling practices (a schedule that seeks to maximize resource utilization, for example, may generate conflict with emergency patient care, leading to the cancellation of one or more elective surgeries), managing patient queues (treating the emergency patients in a hospital on an aggregate basis not distinguishing patients with more severe conditions and that one that should be attended more quickly than others may increase death rates) and resources availability (unavailability of ICU beds or infirmary, equipment, human resources, among others may increase queues, lead to cancellations of surgeries and also increase death rates). They are, therefore, non-random variabilities and must be eliminated.

Considering the stochastic nature of the times between the patients' entrance and the procedure times, the Discrete Events Simulation is used as a tool to allow sizing capacity decision in an operation theater. Discrete Event Simulation (DES) consists in a model that, according to Law & Kelton (2000), can be defined as "a set of assumptions about how a system works, to try to gain some understanding of how the system behaves". For Kelton et al. (2008), this approach refers to the process of designing and creating a computational model of a system to mimic its operations or characteristics in order to better understand its behavior for certain conditions. It is one of the most used tools in Operational Research to identify potential areas of improvement, through the investigation of alternatives from different scenarios in complex systems, such as: military deployment (YILDIRIM et al., 2009), telecommunications systems (HARTANTO et al., 1998), logistic networks (CARVALHO et al., 2012), manufacturing systems (ALDURGHAM & BARGHASH, & PIDD, 2010; KATSALIAKI & MUSTAFEE, 2011), scheduling (DORFMAN & MEDANIC, 2004) e healthcare systems (BRAILSFORD et al., 2009; GUNAL & PIDD, 2010; KATSALIAKI & MUSTAFEE, 2010). In particular, in the healthcare sector, simulation is used in a large variety of studies, including capacity and hospital bed planning (HOLM et al., 2013), intensive care (MALLOR & AZCÁRATE, 2014), operation room configurations and time of waiting (WANG et al., 2009, CHETOUANE et al., 2011), among others. Then, simulation has been used for a number of purposes, since it allows to describe the system in a high degree of detail and to deal with various sources of uncertainty, without simplifying hypotheses that are otherwise necessary to obtain forecasts, like when using other methods, such as mathematical programming or queuing theory (HENDERSON & MASON, 2005). Thus, it is intended from the discrete event simulation model to analyze the trade-off between the scenarios of dedicated rooms availability to the of emergency patients care.

Van Riet & Demeulemeester (2015) categorize types of non-elective patient (trauma, emergent, urgent, semi-urgent, among others) and classify researches founded in different service level (higher waiting time accepted). Categories and waiting time accepted are summarized in Table 1.

In medicine, there are many protocols to establish criterias to define in which severity level the patient disease are and, therefore, which path is necessary to be followed using what set of resources. Thus, hospitals can be classified depending in which specialities and what resources are available. In Brazil, an emergency hospital should be capable to solve all non-elective patient categories except by the trauma (or polytrauma) because the treatment of this kind of disease need many specific

*Table 1. Categorization and waiting time expected by type of non-elective patient (Van Riet and Demeulemeester, 2015)*

| Category | Target (Waiting Time Expected) |
|---|---|
| Trauma | Now |
| Emergent | Since less than 30 minutes to 24 hours |
| Urgent | Since less than 4 hours to 24 hours |
| *Semi*-urgent | Since less than 8 hours to half week |
| Others | No target(fill up free capacity) or since 24 hours to 1 week |

resources and a large variety of health care professionals with rigid coordination rules (SPIEGEL & ASSAD, 2016). According to the same authors, this type of hospital is known as trauma center and is designated to solve only this disease.

## Model Construction and Historical Database

Taking into account the legal requirements pointed out by RDC No. 50 (2002), the board of directors provides one dedicated operation room for emergency patients. Thus, whenever the demand exceeds the attendance capacity of this room (for example, by entering two patients simultaneously or entering a patient during the care of another emergency patient), an elective surgery scheduled must be postponed or canceled.

The actual system identified consists, therefore, in a patient entrance, five operating rooms intended for both emergency and elective patients and one exit. There are three configurations in the model, the first considering all hybrid operation room (the emergency patient is priority, but the room can be used by the elective patient), the second considering one room dedicated to emergency care and the third considering two dedicated rooms.

To construct the conceptual model, field visits and semi-structured interviews were carried out by the authors. It is important to highlight that the model was validated with the interviewees.

A historical database was extracted from the hospital management system. In the database there are the time of entrance of each patient (emergency and elective), the type of procedure performed in the operation theater and duration of the procedure. From these data, those referring to the time of entrance of emergency patients during the hours of the elective patient care (from 7am to 7pm, except weekends and holidays) were extracted and treated, in order to make explicit the competition for resources. The time between arrivals and the surgery duration of emergency patients, both have a random behavior over time, as observed in the historical database. The unit of time used was hours.

The data were processed for further analysis and implementation of the computational model. The demand of elective patients is defined from average of patient admitted rounded down after 10,000 replications without emergency patient entrance.

The following assumptions were adopted in the formulation of the model: (i) it is assumed that the emergency procedure times are independent of the type of procedure (the emergency procedure times were analyzed without considering each type of procedure, due to the inability to predict what type of procedure will be needed per patient); and (ii) it is assumed that all rooms and human resources (teams) have the same operating capacity.

## Discrete Event Simulation

From the data treatment and the adoption of the premises, the samples referring to the period in question were analyzed in order to identify the parametric behavior given by probability distribution curves. For this purpose the data adherence to these curves was tested using the Input Analyzer module of the Arena software. For each random variable of the conceptual model, follow the indications about the distribution pointed out in the literature and the results obtained in the test with the real data in Table 2.

Considering the conceptual model, the results of the adherence tests and the assumptions adopted, the computational model was constructed in the Arena software. First of all, the demand for emergency patients is not taken into account in order to discover the demand of elective patients (the average number of patients that the system can attend without the entrance of emergency patients). To do this, 10,000 replications are performed to find the average number of people that can be allocated on a typical day. The first simulated scenario considers all five-hybrid rooms (with proper priority to the emergency patient in any available room). The second scenario

*Table 2. Adherence test results by Input Analyser*

| Time Between Arrivals | Procedure Time of Emergency Patient | Procedure Time of Elective Patient |
|---|---|---|
| Literature indicates *Exponential* (Van Riet & Demeulemeester, 2015) | Literature indicates *Lognormal, Erlang, Normal, Exponencial* (Van Riet & Demeulemeester, 2015) | Literature indicatesLognormal(Van Riet & Demeulemeester, 2015). |
| Input Analyzer test: *Weibull* | Input Analyzer test:*Weibull* | Input Analyzer test:*Beta* |
| Parameters: -0.001 + WEIB(229, 1.17) | Parameters:16 + WEIB(103, 1.39) | Parameters:30 + 170*BETA (1.24, 1.34) |
| *chi-square p-value:* 0.114 | *chi-square p-value:* 0.42 | *chi-square p-value:* 0,486 |

considers a room dedicated to emergency patient care and the third considers two dedicated rooms. The simulation of the scenarios was performed during the hours of elective patients care (12 hours a day on working days) and replicated 10,000 times.

## SOLUTIONS AND RECOMMENDATIONS

Given the rules and criteria explained in the previous section, the results of the simulation that are discussed throughout this section are consolidated in Table 3.

### Analysis 1: Utilization (Average and Maximum Average) per Scenario

The intention of this analysis is to explain the relation between the utilization rate of the rooms and the dedicated emergency room number. It can be observed that adopting the configuration of all the hybrid rooms (zero dedicated emergency room) increase the average utilization and the quantity of people that can be attended. About the maximum average utilization, it keeps practically constant and it cannot be justified any of the three configurations from this indicator.

### Analysis 2: Emergency Patient Waiting Time (Maximum Value, Maximum Average, and Average) per Scenario

The simulation shows that as the number of rooms dedicated to emergency care increases, the average waiting time of the patient decreases. Therefore, these allowed waiting times should be considered to decide which configuration to adopt.

*Table 3. Arena simulation results considering 10,000 replications*

| | Scenarios (Dedicated Emergency Rooms) | | |
|---|---|---|---|
| | 0 | 1 | 2 |
| Elective demand | 32 | 25 | 19 |
| Utilization (average) | 53,11% | 52,44% | 47,70% |
| Utilization (maximum average) | 84,45% | 84,52% | 82,90% |
| Elective cancellation(maximum average) | 6 | 9 | 6 |
| Elective cancellation(average) | 0,5081 | 0,5734 | 0,5006 |
| Emergency patient waiting time (maximum value) | 2,3978 | 2,06 | 2,1805 |
| Emergency patient waiting time (maximum average) | 1,1108 | 0,7931 | 0,4853 |
| Emergency patient waiting time (average) | 0,3134 | 0,1282 | 0,025766 |

## Analysis 3: Utilization vs. Waiting Time

This is the main trade-off between the three scenarios. From the simulation, it can be clearly seen that the average utilization is inversely proportional to the average, maximum average and maximum value waiting time considering all replications. Thus, it can be concluded that there is no exact answer about the number of rooms that should be dedicated, but the model evidences the system behavior so that a decision is made from the point of view of the level of service that is intended to be delivered (for example, can be based on parameters of waiting time presented in Table 1).

## FUTURE RESEARCH DIRECTIONS

As limitations, rooms and teams were considered like similar, with same capacity and the procedures were treated in an aggregate scale, disregarding the specifics division by medical expertise (SPIEGEL, 2014). For future work, it is proposed to expand this scope, including in the model the other resources involved, which are constraints to the problem, but were not treated in this simulation, as human resources by specialty, equipments, exams, among others. It can be proposed a different set of scenarios, for example, by loading some different number of elective patients that not the average, within the range defined in the first round (maximum and minimum number of elective patients), varying the emergency demand and extending the logic for all the *n* operation rooms with which one wishes to work the Operation Theater.

## CONCLUSION

Given the inherent complexity in the design and management of health units, as highlighted in this chapter, it is worth highlighting the contribution of this text when identifying the pertinence and application of the discrete event simulation approach to evaluate the capacity of a surgical center. Based on their results, it enables the unit's decision makers to position themselves on the possible configurations of the operating rooms, aware of the consequences to the both patient profiles served.

The experiment carried out should be understood as an "alert" to indicate that the postponement or cancellation of elective procedures constitutes a real risk assumed when aiming to increase the utilization rate of operating rooms through the increase in the number of rooms destined to the elective patients attendance and the number of patients *per* day. In view of these conclusions, it is possible, from the proposed method, to challenge the adequacy of the normative instruments in terms of Operation Theater capacity design. In the case studied, the number of rooms currently available to care for emergency patients is not enough to meet this demand immediately.

These contributions are analytical developments of the central contribution of this article, namely: the method used can support the decision to reconfigure the allocation of emergency rooms to the installed capacity of an emergency unit based on historical demand. With this, it increases the degree of understanding about the object, explaining one of the causes of the high number of cancellations of elective surgeries.

## REFERENCES


Abo-Hamad, W., & Arisha, A. (2013). Simulation-based framework to improve patient experience in an emergency department. *European Journal of Operational Research*, *224*(1), 154–166. doi:10.1016/j.ejor.2012.07.028

AlDurgham, M. M., & Barghash, M. A. (2008). A generalised framework for simulation-based decision support for manufacturing. *Production Planning and Control*, *19*(5), 518–534. doi:10.1080/09537280802187626

Brailsford, S. C., Harper, P. R., Patel, B., & Pitt, M. (2009). An analysis of the academic literature on simulation and modelling in health care. *Journal of Simulation*, *3*(3), 130–140. doi:10.1057/jos.2009.10

Brasil. Constituição. (1988). *Constituição da República Federativa do Brasil*. Brasília, DF: Senado Federal: Centro Gráfico.

Carvalho, H., Barroso, A. P., Machado, V. H., Azevedo, S., & Cruz-Machado, V. (2012). Supply chain redesign for resilience using simulation. *Computers & Industrial Engineering*, *62*(1), 329–341. doi:10.1016/j.cie.2011.10.003

Chetouane, F., Barker, K., & Oropeza, A. S. V. (2012). Sensitivity analysis for simulation-based decision making: Application to a hospital emergency service design. *Simulation Modelling Practice and Theory*, *20*(1), 99–111. doi:10.1016/j.simpat.2011.09.003

Dorfman, M. J., & Medanic, J. (2004). Scheduling trains on a railway network using a discrete event model of railway traffic. *Transportation Research Part B: Methodological*, *38*(1), 81–98. doi:10.1016/S0191-2615(03)00006-7

Gino, F., & Pisano, G. (2008). Toward a theory of behavioral operations. *Manufacturing & Service Operations Management: M & SOM*, *10*(4), 676–691. doi:10.1287/msom.1070.0205

Günal, M. M., & Pidd, M. (2010). Discrete event simulation for performance modelling in health care: A review of the literature. *Journal of Simulation*, *4*(1), 42–51. doi:10.1057/jos.2009.25

Hall, R. W. (2013). *Patient flow: reducing delay in healthcare delivery* (Vol. 206). Springer Science & Business Media. doi:10.1007/978-1-4614-9512-3_1

Hart, C. (1998). Doing a literature review: Releasing the social science research imagination. *Sage (Atlanta, Ga.)*.

Hartanto, F., Kreutzer, W., Pawlikowski, K., & Sirisena, H. R. (1996). Quantitative stochastic simulation of telecommunication networks in DESC++. *Computers & Electrical Engineering*, *22*(6), 367–381. doi:10.1016/S0045-7906(96)00015-8

Henderson, S. G., & Mason, A. J. (2005). Ambulance service planning: simulation and data visualisation. In Operations research and health care (pp. 77-102). Springer US. doi:10.1007/1-4020-8066-2_4

Hollnagel, E., Braithwaite, J., & Wears, R. L. (2013). *Epilogue: How to make health care resilient. In Resilient health care* (pp. 227–238). Surrey, UK: Ashgate Publishing Limited.

Holm, L. B., Lurås, H., & Dahl, F. A. (2013). Improving hospital bed utilisation through simulation and optimisation: With application to a 40% increase in patient volume in a Norwegian General Hospital. *International Journal of Medical Informatics*, *82*(2), 80–89. doi:10.1016/j.ijmedinf.2012.05.006 PMID:22698645

Hopp, W. J., & Lovejoy, W. S. (2012). *Hospital operations: Principles of high efficiency health care*. FT Press.

Jacques, J. É. (2007). *Gestão estratégica da criação do conhecimento nas organizações hospitalares: um estudo baseado na construção de protocolos médico-assistenciais*. Academic Press.

Katsaliaki, K., & Mustafee, N. (2011). Applications of simulation within the healthcare context. *The Journal of the Operational Research Society*, *62*(8), 1431–1451. doi:10.1057/jors.2010.20

Kelton, W. D., Sturrock, R. P., Kelton, D. T. D., Sadowski, R. P., & Sturrock, D. T. (2008). *Simulación con software Arena*. McGraw-Hill.

Konrad, R., DeSotto, K., Grocela, A., McAuley, P., Wang, J., Lyons, J., & Bruin, M. (2013). Modeling the impact of changing patient flow processes in an emergency department: Insights from a computer simulation study. *Operations Research for Health Care, 2*(4), 66–74. doi:10.1016/j.orhc.2013.04.001

Lamiri, M., Xie, X., Dolgui, A., & Grimaud, F. (2008). A stochastic model for operating room planning with elective and emergency demand for surgery. *European Journal of Operational Research, 185*(3), 1026–1037. doi:10.1016/j.ejor.2006.02.057

Law, A. M., & Kelton, W. D. (2000). *Simulation Modeling and Analysis*. Singapore: McGraw-Hill.

Litvak, E. (Ed.). (2010). *Managing patient flow in hospitals: Strategies and solutions*. Joint Commission Resources.

Mallor, F., & Azcárate, C. (2014). Combining optimization with simulation to obtain credible models for intensive care units. *Annals of Operations Research, 221*(1), 255–271. doi:10.1007/s10479-011-1035-8

Mantoani, C. C. (2012). *A transfusão sanguínea no perioperatório e a ocorrência da infecção de sítio cirúrgico* (Doctoral dissertation). Universidade de São Paulo.

Marques, I., & Captivo, M. E. (2015). Bicriteria elective surgery scheduling using an evolutionary algorithm. *Operations Research for Health Care, 7*, 14–26. doi:10.1016/j.orhc.2015.07.004

Niu, Q., Peng, Q., El, Y., & Mekkawy, T. (2013). Improvement in the operating room efficiency using Tabu search in simulation. *Business Process Management Journal, 19*(5), 799–818. doi:10.1108/BPMJ-Nov-2011-0081

Paul, J. A., & Lin, L. (2012). Models for improving patient throughput and waiting at hospital emergency departments. *The Journal of Emergency Medicine, 43*(6), 1119–1126. doi:10.1016/j.jemermed.2012.01.063 PMID:22902245

Pedroso, M. C., & Malik, A. M. (2011). As quatro dimensões competitivas da saúde. *Harvard Business Review, 89*, 59–67.

Resolução, RDC (2002). Nº 50 de 21 de fevereiro de 2002. Ministério da saúde.

Shim, J. K., & Siegel, J. G. (1999). *Operations management*. Barron's Educational Series.

Silva, A. C. P. V., Assad, D. B. N., Oliveira, F. L. C., & Spiegel, T. (2016). Uso de simulação de eventos discretos para redimensionamento de capacidade de centro cirúrgico de um hospital de emergência. In XLVIII Simpósio Brasileiro de Pesquisa Operacional - SBPO 2016. Anais do XLVIII Simpósio Brasileiro de Pesquisa Operacional.

Silva, E. L., & Menezes, E. M. (2001). *Metodologia da pesquisa e elaboração de dissertação*. Florianópolis: Laboratório de Ensino a Distância da UFSC.

Slack, N., Chambers, S., & Johnston, R. (2009). *Administração da produção*. Atlas.

Smith, A. C., Barry, R., & Brubaker, C. (2007). *Going lean: Busting barriers to patient flow*. Health Administration Press.

SOMASUS. (2011). *Brasil. Ministério da Saúde. Secretaria-Executiva. Departamento de Economia da Saúde e Desenvolvimento. Programação Arquitetônica de Unidades Funcionais de Saúde / Ministério da Saúde, Secretaria-Executiva, Departamento de Economia da Saúde e Desenvolvimento*. Brasília: Ministério da Saúde.

Spiegel, T. (2013). *Contribuições das ciências cognitivas à gestão de operações: análise do impacto da experiência nas decisões do gestor de operações* (Doctoral dissertation). Federal University of Rio de Janeiro.

Spiegel, T. (2014). An Overview of Cognition Roles in Decision-Making. In Encyclopedia of Business Analytics and Optimization. IGI Global. doi:10.4018/978-1-4666-5202-6.ch008

Spiegel, T., & Assad, D. B. N. (2016). Operations Project and Management in Trauma Centers: The Case of Brazilian Units. In Handbook of Research on Healthcare Administration and Management. IGI Global.

Spiegel, T., & Cameira, R. F. (2016). The Processing Chain of Sterilized Material from the Health Units of Rio De Janeiro County: Analysis and Proposition of Scenarios. In Supply Chain Management: Applications for Manufacturing and Service Industries. New York: Nova Science Publishers.

Spiegel, T., Caulliraux, H. M., Treistman, F. E., Guimaraes, G. E., Vaccaro, G. L. R., Amorim, L. G., . . . Azzam, S. M. (2016). Projeto e Gestão de Operações em Saúde. In Tópicos Emergentes e Desafio Metodológicos em Engenharia de Produção: Casos, Experiências e Proposições. Rio de Janeiro: ABEPRO.

Tieghi, A. L. (2013). *A saúde brasileira tem cura?* Revista Espaço Aberto USP.

Van Riet, C., & Demeulemeester, E. (2015). Trade-offs in operating room planning for electives and emergencies: A review. *Operations Research for Health Care*, *7*, 52–69. doi:10.1016/j.orhc.2015.05.005

Vecina Neto, G., & Malik, A. (2011). *Gestão em Saúde*. Rio de Janeiro: Guanabara Koogan.

Wang, T., Guinet, A., Belaidi, A., & Besombes, B. (2009). Modelling and simulation of emergency services with ARIS and Arena. Case study: The emergency department of Saint Joseph and Saint Luc Hospital. *Production Planning and Control*, *20*(6), 484–495. doi:10.1080/09537280902938605

Xiang, W., Yin, J., & Lim, G. (2015). An ant colony optimization approach for solving an operating room surgery scheduling problem. *Computers & Industrial Engineering*, *85*, 335–345. doi:10.1016/j.cie.2015.04.010

Yıldırım, U. Z., Tansel, B. Ç., & Sabuncuoğlu, İ. (2009). A multi-modal discrete-event simulation model for military deployment. *Simulation Modelling Practice and Theory*, *17*(4), 597–611. doi:10.1016/j.simpat.2008.09.016

## KEY TERMS AND DEFINITIONS

**Elective Surgeries:** The procedure that can be scheduled in advance because there is no imminent risk of death.

**Emergency Department:** A hospital department that may be one of the hospital's entrance doors that is subject to acute peaks triggered by external, and therefore uncontrollable, events where each patient presents a unique set of needs.

**Healthcare Operations Management:** The decisions and actions that occur within the limits defined by the operating system design. These include activities such as the implementation of policies, procedures and strategies, contingent decision-making, process coordination, problem identification and resolution, response to uncertainty and unforeseen problems and rewarding people.

**Hospital Structure:** The structure that fulfill health care, administrative and formative functions to delivery quality health service to the population and it is inserted, by the Brazilian standard, into a health system that is divided into public (SUS) and private, in which the same provider can fulfill functions for both subsystems.

**Non-Elective Surgeries:** Includes emergency or urgent procedures. In both imminent risk of death, but in the first there is not there is no time for preparation nor stabilization.

**Operations Management:** A multidisciplinary field that investigates the design, management and improvement of processes aimed at the development, production, distribution and delivery of products and services.

**Simulation:** The process of designing and creating a computational model of a system to mimic its operations or characteristics in order to better understand its behavior for certain conditions.

# Chapter 7
# Casual Modeling to Improve Customs Services in Iran: A System Dynamics Study

**Alireza Hassanzadeh**
*Tarbiat Modares University, Iran*

**Hoda Eskandari**
*Payame-E-Nour University of Karaj, Iran*

## ABSTRACT

*In its reality, the world is more complex than we expect it to be predicated and controlled; so, by minor view to issues and suppose many phenomena to be independent, it would be impossible to understand their totality and complexity. Utilizing appropriate scientific tools can provide us with useful information to manage decision making to improve methods to carry out operations and apply resources; one of these scientific tools is system dynamics approach. Therefore, based on stages in the methodology of system dynamics and by taking experts comments, in the present study, causal relations between influential factors in improvement of customs service are identified. So, by knowing these causal relations, one can allocate resources and prioritize activities leading to influence on other factors following with the best results. Finally, after validation and having the confirmation of the experts, the researcher has presented a causal model to improve customs services.*

DOI: 10.4018/978-1-5225-3226-2.ch007

## INTRODUCTION

Customs plays an important role in economy, especially that of developing countries. In addition to its financial role in earning money, it has an influential role in the commerce. Further, customs service plays valuable roles in implementation of policies of a country. Since, customs service is a guarantee to price control and domestic production support, existence of a right management pattern will bring about brilliant results in a country. The sensitivity and importance of export of goods and services has led to identifying the tools and factors that are involved and trying to expand exports. The customs service is one of the effective factors in the development of exports of goods and services. Therefore, increasing the quality of customs services is of paramount importance. To improve customs services, several influential factors must have been taken into account. In today's developing and dynamic environment, managers are faced with plenty of variables to make the most appropriate decision or have the best choice. Causal loop diagrams are considered as an important tool to show feedback structure of systems. Having been used for a long time in scientific issues, these diagrams are so common in business world and can be considered as the best alternative to rapidly obtain a hypothesis on the reasons of dynamism, deduction and understanding mental models of individuals and groups; further, they are appropriate tools to relate important feedbacks which play a role in creating the problem.

In most of the developed countries, customs mechanism depends upon efficient use of human resources and equipment to improve services. To find a way to form the truth as a system instead of a series of independent issues would be necessary to make a suitable plan. Any system is defined according to collectivity and interaction among its constituents. To improve a system, its totality should be modified as a whole, not each part, separately. Because improvement of constituents of a system may not lead to improvement of its totality, it even can be worse (Ahmedvand & Arab, 2009).

The system dynamics approach provides holistic view. This approach was introduced in early 1960s by Forster at MIT University (Sterman, 2008). It is a branch of system thinking which studies totality of a system without tangling itself with ambiguous and complex terminology. System dynamics is a method to study and manage complex systems with feedback. The reason for using system dynamics is having a comprehensive view of improving customs services. Traditional techniques ignore the dynamics of customs services improvement and can't understand the complexity of the issues. Therefore in this chapter with a system dynamics approach a casual modeling are provided to improve customs services.

# BACKGROUND

## Service Improvement

'Service' is meant a desired result by costumers, it is an activity or benefit supplied by one side for another; it is basically intangible with no sense of possession. This result may be physical or immaterial (Kotler & Armstraong, 2009). Though excellent service is a winning strategy, many organizations waste their money and energy to improve service by incomprehensible plans as well as management bluffs instead of taking practical measures (L. Berry,A. Parasuraman, 1994).

Increasing competition, which has been started in 1950s and has been intensifying significantly, brings about the need to adopt modern approaches to make improvements. After World War II, applying many of improvement approaches has been increased; these approaches continue to be used. It was in Japan where using tools and techniques for improvement got started and spread; including, prevention, Total Productive Maintenance (TPM), Total Quality Management (TQM) which is invented by Juran and Deming. Japanese sub procedures are Kaizen, 5s and lean-time. These concepts have been transferred to the west causing some countermeasures to Japanese; such as Theory of Constraints (TOC), Business Process Reengineering (BRP), Business Process Improvement (BPI) and other approaches related to Kaizen. Many of these approaches have no clear solution or guidance regarding decision-making for improvement of different constituents; furthermore, little guidance exists to evaluate them. In the literature of the present study, there are a lot of processes and methods to evaluate them; however, most of them are static, while, flexible and dynamic processes require flexible approaches (Grünberg, 2004).

## Improvement of Customs Services

Since more than a decade, customs offices are faced with many pressures and various demands from political institutions, law enforcement agencies and business communities around the world; such as to modernize customs service in line with development of information technology and world commerce; besides, some challenges arose such as progress in commercial facilitation, national security and income collection (McLinden,2005). In custom offices of many countries, much effort is done to decrease corruption, facilitate commercial activities and increase revenues. However, there is a little literature and there is no comprehensive study in this area world-wide (Cantens, Raballand & Bilangna, 2010).

Evaluation of service improvement can be performed in various forms; however, because of having a vast range of clients regarding customs services, recognizing vital aspects of approaches and processes, decision-making based on facts and determination of requirements for improvement of services increases the importance of this issue (McLinden, 2005). Some studies and activities carried out in this area around the world are as follows:

Balanced Scorecard (BSC), Key Performance Indicators (KPI) which are used in customs offices of United Kingdom to be assured in being in a right way to their objective (McLinden,2005), Economic Value Added (EVA), Activity-Based Costing (ABC), Total Quality Management (TQM), staff training (Ucbasaran,Westhead & Wright,2006), investing in human resources (Chandler, G & Hanks, 1998), coordination and improvement of approaches (Wulf L De, 2004), modernization of customs offices and electronic customs (Apples and Struye de Swieland, 1998; GwardzińskaEwa, 2012), campaigning against corruption in offices, to increase security and to promote controlling equipment (Wulf L De, 2004), risk management (Geourjon, A-M., 2005) and utilization of data processing for customs management (Bertrand Laporte, 2012), single window (Luciano Pugliatti, 2011), proper management of resources (Michael Haughton & Sapna Isotupa, 2009), the proper time to receive shipments (Shingo Matsuda, 2012), electronic management of boundaries (Holloway, 2009), quality management (YangYu, 2012), x-ray photography installations, registration and review of vehicles number plate, vehicles tracing, update and equip systems to good's clearance, approaches to present declaration, testing and clearance of goods in ports, structural improvements, reengineering of customs process, etc.

## System Dynamics

Richmond (1994) defines systematic thinking as "art and science of join structures to performance and performance to structure; often to change structures (communications) to improve performances." To understand a phenomenon, one should study them from a dynamic point of view, so, it will be possible to identify changes. These changes are created because of causal relations among many factors. According to causal philosophy, there is always a cause for any effect (Sushil, 1993).

The approach of system dynamics was introduced in early 1960s by Forster at MIT University (Sterman, 2008). It is a branch of system thinking which studies totality of a system without tangling itself with ambiguous and complex terminology. System dynamics is a method to study and manage complex systems with feedback.

## Causal Loop Diagram

Compiling and drawing causal circles of a system is considered as one of the most important stages in designing and analysis of systems related to dynamic phenomena.

This diagram illustrates relations which are considered as causal cyclic chains in real systems.

## MODELING

In modeling process, the model maker defines a limit for the model to internalize causal framework. The main purpose of methodology of system dynamics is to know and have a deep insight on system's function which contributes to design improved policies to perform the said system (Sushil, 2008).

### Modeling Stages of System Dynamics

The process of modeling starts with problem defining. By observing a phenomenon as a totality and deriving its reference behavior which clarifies changes in the past, model maker can define some concepts as mental models; by the help of systematic knowledge and data obtained from experts as well as theories of the matter under research, mental models can be expressed as descriptive models. These statements are dynamics hypotheses. After suggesting hypotheses, the main variables are identifies within dependent and independent variables along with their relationship and how they affect one another resulting in formation of causal loops (Doroodian, 2008). The process of extracting mental model and changing it to causal relations based on system dynamics is as follows:

1. To express system's scope
2. To extract and compile a dynamic hypothesis
3. To build and present causal model
4. To confirm the model by experts

### Problem Definition (Expression of System's Scope)

Problem and subject of modeling in the present research is the relation between influential factors in improvement of customs services. The purpose of this causal model is to show these relations; i.e. clarifying this model can help policymakers and managers to identify Priority of influential factors in improvement of a specific area and have a certain plan for that. Further, this model can be applied as a tool for evaluation and troubleshooting.

Because this model is provided to improve customs service, its scope includes three groups of customs service as receivers of customs (staff, tradesmen and the government); then effective factors in improvement of these services are examined.

## Operating Definition of Variables

Variables of staff sector:

In this part, variables consist of factors which have a great influence on improvement of services to customs staff.

- Staff training
- Salary, insurance and benefits
- Facilities for staff
- Evaluation system for employees

Variables of services to traders:

Variables of this part include factors which have a great influence on improvement of services to traders.

- The amount of time needed to perform a customs process
- The time at which tradesmen have accessibility to necessary information
- The cost of visiting for tradesmen
- The quality of services
- Improvement of services to customs personnel

Variables of government sector:

Variables of this part include factors which have a great influence on improvement of services to the government.

- Custom duties
- Severity of violation
- Information systems
- Control and enforcement of policies
- The total period of performance

## Extracting and Compiling Dynamic Hypothesis

Causal loop starts at accurate allocation of resources. By categorizing customs service in three groups of receivers of services, some of the relations between elements are expressed as follows:

Increase in salaries and benefits of staff tend to increase in facilities and welfare programs for them, and increase budget for their training. Existence of an appropriate evaluating system results in improve services for personnel. Improve services to custom staff, allocate more budgets to improve quality of services, decrease the access time to information for tradesmen, decrease the costs for tradesmen to refer to customs offices in person, decrease the time needed for all the process which leads to improvement of services to tradesmen. Furthermore increased the budget allocated to reengineering, design and implement necessary software and hardware infrastructures will result in: decreased the period needed to access to information by the tradesmen, decreased costs needed for tradesmen to refer in person, decreased the total time of the process, eliminated redundant processes, etc.

Decrease in total time for customs process cause an increase in tradesmen's benefit. Elimination of manual and redundant processes and has them done electronically leads to decrease in human errors and costs of the process performance.

Increase in custom duties, decrease of the time needed to implement the process, development of information systems, decrease in costs and the period of time needed to reporting, providing statistics and information, increase in quality control, increased ability to control, enforce policies and decrease violations all of which lead to improvement of services received by the government.

## To Obtain Main Variables of the Model via Asking Experts

To determine main variables, a primary list of suggested factors and variables and their relations for three parts of staff, tradesmen and the government are compiled and are distributed in the form of a questionnaire among 11 academic and experienced experts in customs. They had to declare their agreement or disagreement for each variable along with its relationships. In the questionnaires, the Likert continuum was applied (totally agree, agree, no opinion, disagree, and totally disagree). Analysis of the mentioned questionnaire has been done utilizing Sign test.

## Causal Diagrams

After determining dynamic hypothesis, important causal circles can be drawn. Compiling and drawing causal cycles of a system is one of the stages in design and analysis of systems related to dynamic phenomena. The causal diagram is drawn using Vensim software.

In Figure 1, causal model related to improvement of customs service is shown which includes improvement of services to three groups of service receivers including customs staff, tradesmen and governments:

The causal model related to service improvement to staff is shown in Figure 2.

The causal model related to improvement of services to tradesmen is shown in Figure 3.

The causal model related to improvement of services to government is shown in Figure 4.

*Figure 1. Causal relations of service improvements*

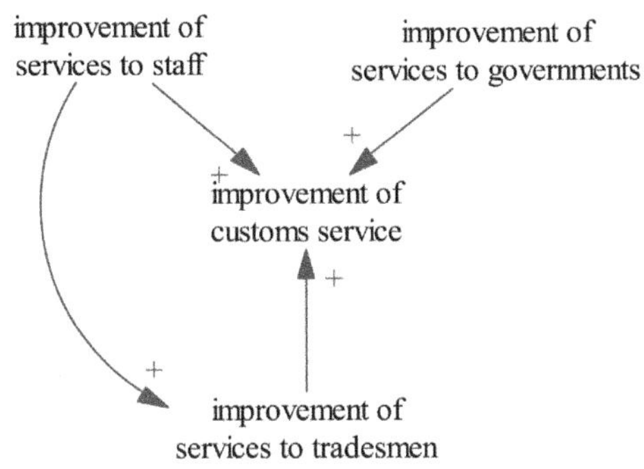


*Figure 2. Causal relations to improve services to customs staff*

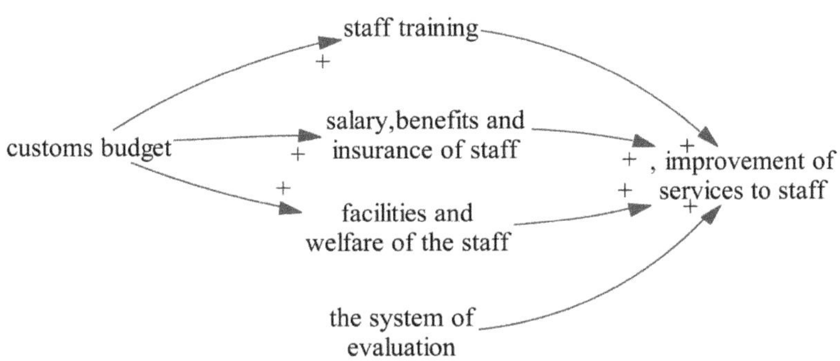


*Figure 3. Causal relations of customs service improvement to tradesmen*

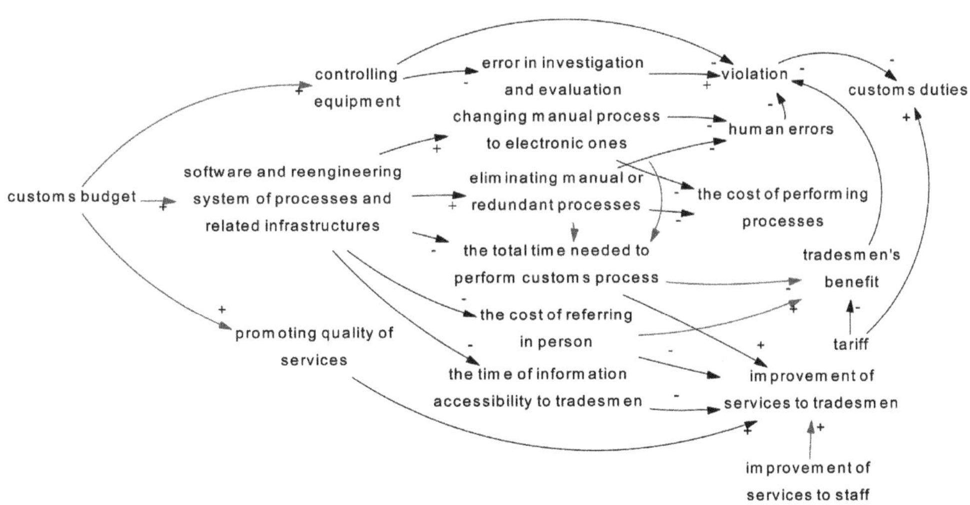

*Figure 4. Causal relations of customs service improvement to government*

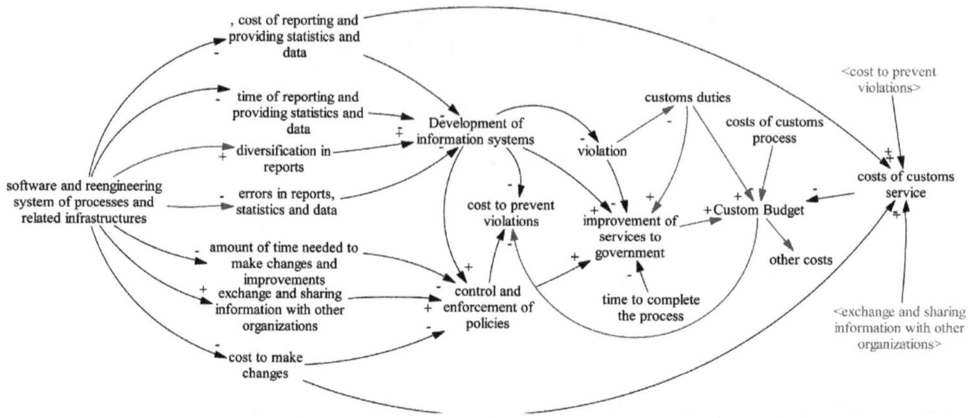


In these diagrams, causal interactions inside the system are illustrated regarding the direction and manner of their influence.

As it is shown in figures above, by using dynamic systems, one can show effective factors in customs service improvement as a series of causal relations whose effect on one another is shown by + and – signs. Due to variety of customs services, these diagrams are categorized and shown as a model in three groups, customs staff, tradesmen and the government. The mentioned factors affect improvement of customs service directly or indirectly whose relations and the manner of influencing on one another are shown in causal diagrams.

## Validation of the Model and Experts' Confirmation

As it has been explained, to validate factors and variables, questionnaires of effective factors and relations in improvement of customs service have been distributed among academic experts. After analyzing their responses, the researcher has omitted some factors and made required changes. Then, based on these diagrams and by interviewing with five experts in system dynamics ask them to comment regarding existence of this causal relations and their effect on principal defined performing variables. The general causal diagram for customs service improvement is drawn in Figure 5 after approved by experts.

As it is shown in Figure 5, there is a causal relationship among important and key variables. There are repetitive mutual relations among variables; so, each of them has an influence on others in dynamic, causal relationships.

*Figure 5. Causal model to improve customs services*

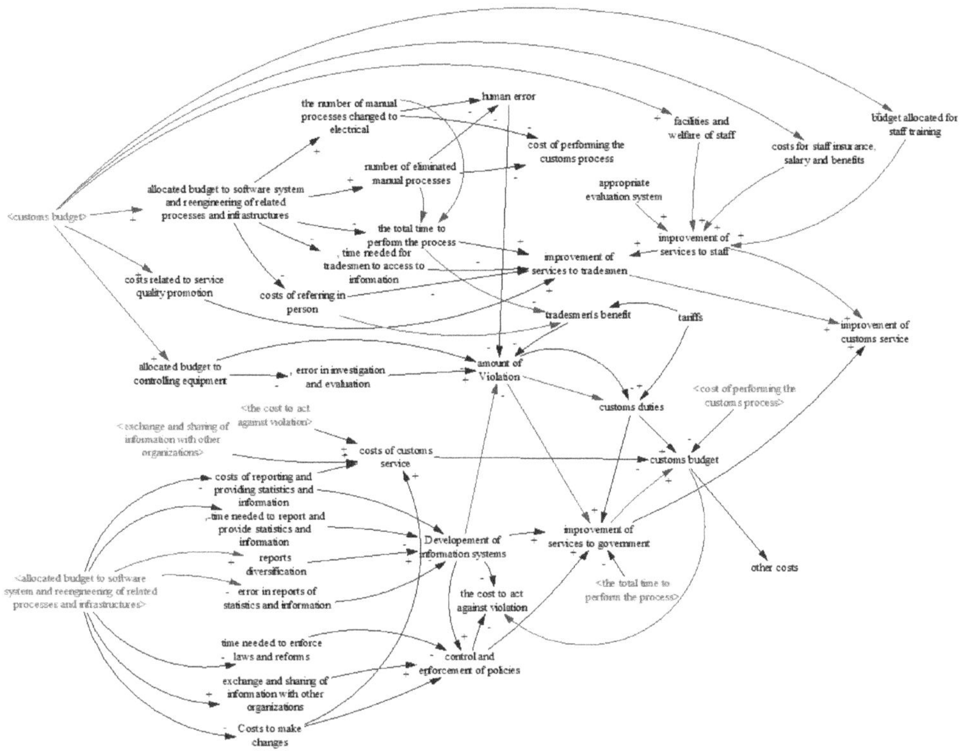


## CONCLUSION

The complexity of the problems of today's societies, requires new analytical tools and techniques. Causal diagrams are powerful tools to draw feedback infrastructure of complex systems. In the present research, a brief review of measures rendered related to improvement of customs services and concepts, modeling steps of dynamics systems methodology has been done through conducting interview and distribution of questionnaire and obtaining a mental model of experts and a dynamic hypothesis has been compiled. In this process, main variables are extracted through distribution of questionnaire among experts and causal circles have been drawn. After presenting that model to the experts of system dynamics, finally, general suggested model has been drawn as a series of causal relations whose influence on each other has been shown by plus(+) and minus(-) signs. In this general model, several causal circles have been observed which create circled relations in the system leading to a definition of the dynamic model. These diagrams provide a comprehensive view for the decision makers. Based on the proposed model, managers and practitioners can make better decisions for improvement of customs services.

## REFERENCES

Ahmedvand & Arab. (2009). *Application of system dynamics methodology in police*. Tehran: Organizational Development of Police.

Berry, L. L., Parasuraman, A., & Zeithaml, V. A. (1994). Improving Service Quality in America. *The Academy of Management Executive Journal of Marketing, 8*(2), 31-46.

Cantens, T., Raballand, G., & Bilangna, S. (2010). *Reforming Customs by measuring performance: a Cameroon case study*. World Customs Journal.

Doroodian, H. (2008). *Applying Dynamic Systems in Balanced Scorecard to Design Plans of Dynamic Strategies of Small and Medium Industries of Iran* (MS Thesis). Science and Technology University, Tehran, Iran.

Grünberg, T. (2004). Performance improvement Towards a method for finding and prioritising potential performance improvement areas in manufacturing operations, Management. *Journal of Productivity and Performance International, 53*(1), 52–71. doi:10.1108/17410400410509969

Kotler, P., & Armstrong, G. (2009). *Principles of Marketing*. Prentice Hall.

McLinden, G. (2005). *Integrity in Customs, Customs modernization handbook*. Washington, DC: World Bank.

Sterman, J. D. (2000). *Business Dynamics: Systems Thinking and Modeling for a Complex World*. McGraw-Hill/Irwin.

Sushil. (1993). *System Dynamics: A Practical Approach For Managerial Problems*. Wiley Eastern Ltd.

Ucbasaran, D., Westhead, P., & Wright, M. (2006). Opportunity identification and pursuit: Does an entrepreneur's human capital matter? *3rd Specialized Research Workshop, Gate2*. Growth Academic Network, University of Amsterdam.

# Chapter 8
# A Model for the Establishment and Development of Telework in Public Organizations:
## Using the System Dynamics Approach

**Alireza Hassanzadeh**
*Tarbiat Modares University, Iran*

**Robabeh Sadat Emami Meybodi**
*Payame-E-Nour Unicersity of Karaj, Iran*

## ABSTRACT

*In this chapter, based on systems dynamics approach, efforts have been made to provide a model for the establishment and development of telework in public organizations of the Iran. The main goal of the study is to find a suitable and practical model for the establishment and successful development of telework in the public organizations of the country. Modeling for the establishment and development of telework for the first time in this research has been done using the systems dynamics approach. One of the main findings of this study is the extraction of the factors and main variables that are effective in the establishment and development of telework. Also, the effect and the relationship between the factors and operational variables of the model were revealed and the components of the causal loop were confirmed. Finally, the final model of the research was presented.*

DOI: 10.4018/978-1-5225-3226-2.ch008

## INTRODUCTION

The traditional patterns of work, workplace, work time, organization, and employer's attitudes are changing because of the rapid advancement of information and communication technologies. Teleworking is a product of the rapid expansion of ICT. System dynamics is a methodology for studying, analyzing, simulating and improving socioeconomic and managerial systems using a feedback approach. Experience has shown that using system dynamics simulation improve decision making quality.

This research intends to introduce a dynamic system model that considers the most important variables affecting telework in order to establish and develop it in the country. In other words, this research seeks to find a suitable and practical model for the successful establishment and development of telework in the public organizations of the country. The resulting model can provide managers with a comprehensive view to make effective decisions in this regard.

## LITERATURE REVIEW

Since the late 1970s, when microcomputers and modems became available to the public, the dream of "teleworking" was fully realized so that its occurrence caused a significant leap in productivity.

Jack Nilles, who is mentioned in various books as the father of telecommuting, was the first to formally investigate the issue and the telework idea was raised for the first time in the 1970s. In Nils' opinion, people do not have to go to work in the future, but work will come to them (ziari, 2009).

The telework was actually launched in 1973 during a California air pollution control project (Khani, 2011). And in Europe, teleworking since 1980, after prevalence the unemployment rate, has begun. Futurists then predicted that telework would be the dominant form of organizing work in the future. Among these people is the famous "Alvin Toffler" who presented the idea of an "electronic cottage" (Burch, 1991). In the 1980s, there was a lot of research in Europe about the positive and negative aspects of telework, as well as job design and organizational structure (Huws et al., 1990).

In 1987, Francis Kinsman, who had more than 20 years of futuristic experience, published a book titled " Telecommuters". This is the first book in the UK to deal with the predictability of telework and the revolution of "flexible work" (Sohrabi, 1382). In 1995, the European Commission announced that telework might be part of the future work of the people. So, as Nilles predicted, Teleworking is being implemented in many public and private organizations around the world and they benefit from it (Hill and Alan, 1996). The governments and large industrial companies in the world who seek to increase productivity and enhance the quality of life of their employees have welcomed this phenomenon and have gained a lot of benefits for themselves and their employees. However, the teleworking average in the world is less than 5%, and only in the United States, due to the existence of computer companies with their philosophy of work, has a great deal of teleworking compatibility (Jalali, 2010).

Bernardino et al. (1993) proposed a comprehensive framework for telework, which could be used as a means of predicting the use of telework.

Sullivan et al. (1993) utilized priority data of employees in three different US cities to model the participation rate of employees in telecommuting. One of the objectives of this study were to identify the determinants of the employee's decision to choose telecommuting. The models were made of three variables: travel characteristics, occupational characteristics, and socioeconomic characteristics of the decision maker.

Mokhtarian and Salmon (1995) in a research on a statistical sample of 628 employees sought to identify barriers and facilitators of teleworking. To this end, a logit model was used to identify the desire for telework from home. With the awareness of the fact that organizations have the right to compel telecommuting or delegate it to their employees, Belanger examines the conditions of telework selection by employees.

## RESEARCH METHOD

The research method of this study is a descriptive-survey. The statistical population of this study consists of two groups of people:

1. Academic experts who have expertise in computer science and have some research in this field, as well as experts in the field of system dynamics.
2. Organizational or industrial experts that include those who have the responsibility of information technology and virtualization in the government organizations, as well as the staff and management of the teleworking department of the public organization.

The system dynamism approach is used to model the establishment and development of telework in public organizations. There are several softwares available to simulate dynamic systems. In this research, Vensim software is used for modeling.

## MODELING THE ESTABLISHMENT AND DEVELOPMENT OF TELEWORK IN PUBLIC ORGANIZATIONS

Extracting the mental model and transforming it into cause and effect relationships according to the dynamics of systems for the establishment and development of telework in governmental organizations are as follows:

### Identify and Define the Problem

Teleworking as a new way of working in organizations will bring many benefits for the society and organizations, and will solve many of the problems that today threaten the metropolis. Of course, all the benefits of teleworking depend on successful implementation and development in the public organizations of the country, and also depends on the recognition of all the factors involved in the establishment of telework and the effective variables in its development. It is only by systematically looking at this phenomenon that all the affecting factors can be identified and reviewed. Therefore, the purpose of this research is to design a model that describes the variables and factors associated with the establishment and development of telework in order to better understand this emerging phenomenon and help design appropriate strategies for its successful establishment and development.

### Statement of the System Boundary

This stage is in fact the most important stage of modeling and states that what is in the scope of modeling, and what is not. The issue of modeling in this research is to find the relationship between the factors affecting the establishment and development of telework. Therefore, the relationship between the factors and the variables that are effective in telework establishment, as well as the causes and factors that lead to telework development, should be examined in a systematic model.

In fact, the clarification of this mental model is intended to help policy makers and other public agents to better understand the effective factors in telework establishment and development and know if they plan to implement or develop a telework plan in an organization, they should pay attention to all relevant factors in order to succeed. In this modeling, we try to consider all the factors and variables that somehow affect the establishment of telework and its development in organizations. Therefore, the scope of modelling in this research are limited to the factors involved in the establishment and development of telework.

## Extracting Performance Criteria and Compiling Dynamic Hypotheses and Drawing the Initial Model of Research

To provide causal relationship models, factors should be defined for each of the studied domains. Meanwhile, these factors must be converted into operational variables in order to illustrate the cause and effect structure, in a way that is operational and measurable in an organizational environment.

### Extracting Factors

In order to design a research model, all the criteria in the theoretical literature were carefully studied and extracted. Then some of them were removed using the opinion of the academic experts.

### Formulating of Dynamic Hypotheses

Based on the previous studies, the relationships between factors can be expressed as follows which are in fact the initial hypotheses of the research:

The starting point of the casual loop is the amount of investment on the infrastructure of information and communication technology. Increasing in these investments will increase the number of remote work centers in the country. As the number of these remote work is increased, which will provide telework facilities for teleworkers, the telework establishment in the country will increase.

On the other hand, increasing the amount of investment on ICT infrastructure will strengthen the security of information networks of organizations that in turn leads to more teleworking. Increasing the amount of investment on information and communication technology infrastructure will increase the accuracy of teleworkers' performance evaluation systems that in turn leads to a more accurate measurement of the teleworker's outcomes and ultimately leads to increased telecommuting.

Increasing investment in teleworkers education has led to the adoption of a telework culture in the organization, which will increase the telework establishment. The development of telecommuting rules and regulations and the increase of the rules in this area will increase the telework establishment. With increasing teleworking in organizations, organizational overhead costs (public spending such as electricity, gas, etc.) are reduced. Following the reduction of these additional costs, the efficiency of the organization increases, that leads to there will be an increase in the efficiency of the increase in productivity. By increasing productivity, the amount of telework attractiveness increases, which in turn raising the establishment and development of teleworking.

With increasing in telework establishment, the amount of employee absenteeism decreases. Reducing absenteeism leads to an increase in organizational effectiveness and productivity. Increasing telecommuting will help to those who can not work for certain problems, such as some women or the disabled, find employment opportunity in remote situations.

Employment growth will lead to lower unemployment. Reducing unemployment will also result in economic growth. By increasing this growth and economic development, the amount of telework attraction will be increased.

With increasing in employment, immigration to other cities is reduced, because in the telework conditions, employment opportunities are available to everyone in all parts of the country. As a result, with the reduction of immigration, the growth rate of big cities is decreasing. Reducing population growth in large cities will reduce traffic and traffic volumes. Increasing telecommuting will reduce traffic volume, because it is no longer necessary for employees to go to work or return home during certain hours of the day. By reducing traffic, fuel and energy consumption decreases. Reducing environmental pollution reduces fuel consumption and energy consumption. Following the reduction of environmental pollution, the amount of telework attractiveness increases.

With the reduction of traffic volume, employees's waste time decreases. It is no longer necessary for employees to spend their valuable time behind heavy traffic. They can save this time spent on work affairs, thus increasing the effectiveness of the organization. Reducing traffic will reduce individual staff costs. Reducing individual costs will increase the level of social welfare and therefore increasing the attractiveness of telework. On the other hand, with the reduction of traffic volume, the cost of infrastructure and the development and maintenance of roads will be reduced. By reducing these costs, the amount of economic investment will increase, which leads to an increase in economic growth.

Increasing telework reduces demand for buildings and land. This decline in demand leads to lower prices. It leads to lower inflation and ultimately, it will raise the level of social welfare. With the rise of telework, the level of social security is rising. Subsequently, it reduces the cost of providing this security to the community. The result will be an increase in the amount of investment in the economy.

## Drawing the Causal Model Relationships and Constructing and Presenting the Cause and Effect Diagram (Initial Model of the Research)

By identifying the dynamic hypotheses, important causes and effects loops are created. This diagram represents the relationships that are referred to as real-world causes and effects loops. Initially, in order to form the initial hypotheses, based on the literature study, the mental model of the subject was extracted. Its cause and effect loops are drawn based on the system dynamics method. According to the research subject, the cause and effect model of the two areas of telework establishment and telework development were presented separately using a systematic approach. Finally, the initial model of research was drawn up by integrating these two models.

## Extracting the Main Variables of the Model and Modifying the Initial Model and Evaluating It by the Experts

In order to extract the main variables of the model, these factors extracted in the form of a questionnaire were provided to academic experts and field experts. After collecting their comments, the main variables related to telework establishment and development were extracted as shown in Table 1.

## Modify the Initial Model

Finally, the initial model of the research was modified based on the expert opinions. In this way, with the approval of experts, in addition to extraction of the main variables of research topic (establishment and development of telework), the main dynamic hypotheses of the research model were also extracted. The final model of the research is presented in two areas of telework establishment and telework development. Finally, the final model otained as presented in Figure 1.

In this diagram, causal relationships are identified within the system with regard to their direction and effects. As it is evident from the figure, with the use of dynamic systems, the modelof establishment and development of telework has been a series

*Table 1. Main variables that affects on the establishment and development of teleworking*

| Operational Variables for Telework Establishment | Operational Variables in the Field of Telework Development | |
|---|---|---|
| • The amount of investment on information and communication technology infrastructure<br>• Number of remote work centers<br>• Security of Information Networks of the Organization<br>• The accuracy of teleworkers performance evaluation systems<br>• The amount of investment in teleworker training | • Organizational overhead costs<br>• Efficiency<br>• Productivity<br>• The amount of absenteeism<br>• Effectiveness<br>• Employment rate (creating jobs)<br>• Unemployment rate<br>• The rate of Economic growth and development<br>• Immigration rate | • The growth rate of big cities<br>• Time waste<br>• Traffic volume and traffic<br>• Fuel and energy consumption<br>• Rate of Environmental pollution<br>• Individual costs<br>• The level of social welfare<br>• The amount of telework charm |

*Figure 1. The final model of research for the establishment and development of telework*

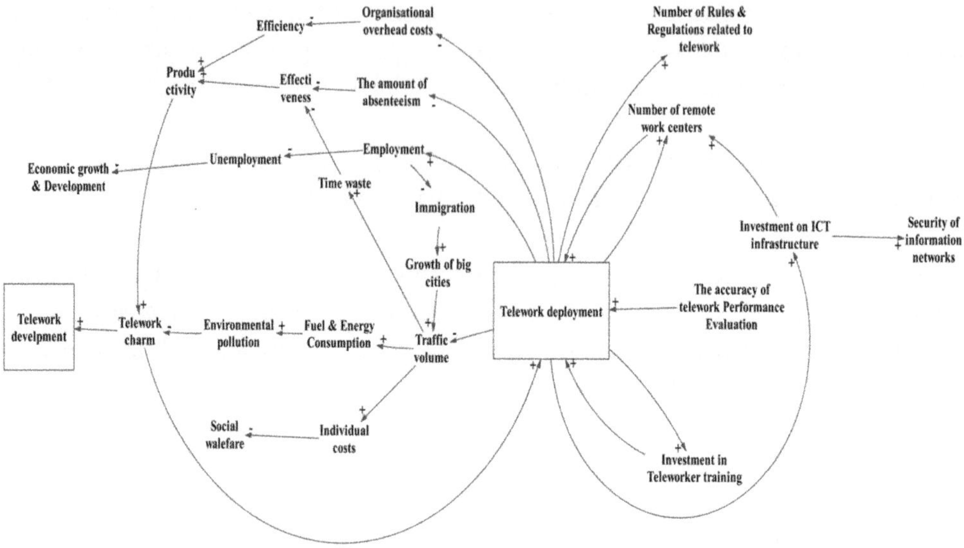


of cause-and-effect relationships which their impacts shown by + or – signs. Each of them is effective in decreasing and increasing another dynamically and in the form of causal relationships. Of course, these effects are carried out according to the concepts of dynamic systems with a reasonable delay. Feedback loops represent the dynamics of the model.

## DISCUSSION AND CONCLUSION

The advancement of information and communication technology has changed the type and way of human life and even how it works. Now individuals and corporations, through new technologies can do activities and operations that were previously impossible to do. The Internet and, consequently, e-commerce, in turn, are a phenomenon derived from information technology that made a profound transformation in communications and information transfer processes. Now people are able to do their work in the home without a physical presence at their workplace, using only a personal computer and a modem.

Telework have been considered with the aim of increasing the efficiency of workr force as well as the effectiveness and efficiency of organizations in our world and our country. The subject of remote work in Iran has been seriously raised, and the Government of the Islamic Republic of Iran has tried to reach the objectives of the telework program, which is one of the ten most important aspects of the administrative transformation and the latest achievement of modern administrative systems. Organizations that have some kind of telework ability will benefit from the numerous results of the above mentioned project. Therefore, in order to better understand such issues as telework in organizations, using a dynamic system modeling approach, a telework conceptual model should be designed that shows the important variables in this field.

Considering the need to design an applied model for the establishment and development of telework in organizations, in this research, it has been attempted to design a conceptual model with the help of system dynamics approach and drawing of causal diagrams. It includes the most important and influential factors related to telework, with a view to its successful establishment in the country and its further development in the public organizations of the country.

## REFERENCES


Barlas, Y., & Yasarcan, H. (2006). Goal setting, evaluation, learning and revision: A dynamic modeling approach. *Evaluation and Program Planning*, 29.

Baruch, Y., & Yuen, Y. K. J. (2000). Inclination to opt for teleworking. *International Journal of Manpower*, *21*(7), 521–539. doi:10.1108/01437720010378980

Brocklehurst, M. (2001). Power, identity and new technology in homework: Implications for new forms of organizing. *Organization Studies*, *22*(3), 445–466. doi:10.1177/0170840601223003

Burch, S. (1991). *Teleworking: a Strategic Guide for Management*. London: Kogan Page.

Fireman, S. M. (1999). *A model of telecommuting withdrawal: Employee perceptions predicting the reduction or stopping of telework* (Ph.D. diss.). University of Washington.

Forrester, J. W. (1961). Industrial Dynamics. Cambridge, MA:MIT Press.

Gray, M., Hodson, N., & Gordon, G. (1994). *Teleworking Explained*. Chichester, UK: John Wiley & sons.

Handy, S. L., & Mokhtarian, P. L. (1996). The future of telecommuting. *Future*, *28*(3), 227–240. doi:10.1016/0016-3287(96)00003-1

Hill & Hawkins. (1996). Work and family in the virtual office. *Family Relations, 45*(3).

Huws, U., Korte, W. B., & Robinson, S. (1990). Telework: Towards the Elusive Office. Chichester, UK: John Wiley & Sons.

Illeges, V., & Verbeke, A. (2004). Telework: What dose it mean for management? *Long Range Planning*, 32.

Jalali, A. A. (2010). Telework, Should and should not. *The Journal of Information Era Analysts.*

Kelly, M. M. (1985). The next workplace revolution: Telecommuting. *Supervisory Management, 30*(10).

Khani, M. (2011). *Developed Administrative System and Teleworking*. Modir Site.

Kowalski, K. B., & Swanson, J. A. (2005). Critical success factors in developing teleworking programs, Benchmarking. *International Journal (Toronto, Ont.), 12*(3).

Lister, K., & Harnish, T. (2010). *Work shifting benefits: The bottem line*. Telework Research Network/Aerodite Inc.

Moorcreft & Bennett. (1995). *European Guide to Teleworking: A Frame Work for Action.* Dublin, Ireland: European Foundation for the Improvement of Living and Working Conditions.

Nilles, J. M. (1998). *Managing Telework: Strategies for Managing the Virtual Workforce*. New York: John Wiely & Sons, Inc.

Perez, M. (2004). A technology acceptance model of innovation adoption: The case of teleworking. *European Journal of Innovation Management, 7*(4), 280–291. doi:10.1108/14601060410565038

Perez, M., Sanchez, A. M., Carnicer, M. P. L., & Jimienez, M. J. V. (2004). The environmental impacts of telewoking: A model of urban analysis and a case study. *Management of Environmental Quality, 15*(6). doi:10.1108/14777830410560728

Pierson, K. M. (2006). *Home-based work: A form of telework with advantages and disadvantages related to health outcomes* (Master's thesis). Northern Kentucky University.

Pophal, L. G. (2002). *Telecommuting: Managing off-site staff for small business.* International Self-Counsel.

Ruth, S., & Chaudhry, I. (2008). *Telework: A productivity paradox? Public Policy.* IEEE Computer Society.

Siha, S. M., & Monroe, R. W. (2006). Telecommuting's past and future: A literature review and research agenda. *Business Process Management Journal*, *12*(4), 455–482. doi:10.1108/14637150610678078

Sohrabi, B. (2003). *Teleworking, Concepts and Planning Process*. Tehran: Irandoc.

Spinks, A. W. (2001). Satellite and resort office in Japan. *Transportation, 18*.

Sterman, J. D. (2002). All Models Are Wrong: Reflections on Becoming a Systems Scientist. *System Dynamics Review*, *18*(40).

Sushil. (1993). *Systems Dynamics: A Practical Approach for Managerial Problems*. Wiley Eastern Limited.

Ward, N., & Shabha, G. (2001). Teleworking: An assessment of socio-psychological factors. Facilities, 19(1-2).

# Related References

To continue our tradition of advancing academic research, we have compiled a list of recommended IGI Global readings. These references will provide additional information and guidance to further enrich your knowledge and assist you with your own research and future publications.

Adeyemo, O. (2013). The nationwide health information network: A biometric approach to prevent medical identity theft. In *User-driven healthcare: Concepts, methodologies, tools, and applications* (pp. 1636–1649). Hershey, PA: IGI Global. doi:10.4018/978-1-4666-2770-3.ch081

Adler, M., & Henman, P. (2009). Justice beyond the courts: The implications of computerisation for procedural justice in social security. In A. Martínez & P. Abat (Eds.), *E-justice: Using information communication technologies in the court system* (pp. 65–86). Hershey, PA: IGI Global. doi:10.4018/978-1-59904-998-4.ch005

Aflalo, E., & Gabay, E. (2013). An information system for coping with student dropout. In L. Tomei (Ed.), *Learning tools and teaching approaches through ICT advancements* (pp. 176–187). Hershey, PA: IGI Global. doi:10.4018/978-1-4666-2017-9.ch016

Ahmed, M. A., Janssen, M., & van den Hoven, J. (2012). Value sensitive transfer (VST) of systems among countries: Towards a framework. *International Journal of Electronic Government Research, 8*(1), 26–42. doi:10.4018/jegr.2012010102

Aikins, S. K. (2008). Issues and trends in internet-based citizen participation. In G. Garson & M. Khosrow-Pour (Eds.), *Handbook of research on public information technology* (pp. 31–40). Hershey, PA: IGI Global. doi:10.4018/978-1-59904-857-4.ch004

Aikins, S. K. (2009). A comparative study of municipal adoption of internet-based citizen participation. In C. Reddick (Ed.), *Handbook of research on strategies for local e-government adoption and implementation: Comparative studies* (pp. 206–230). Hershey, PA: IGI Global. doi:10.4018/978-1-60566-282-4.ch011

Aikins, S. K. (2012). Improving e-government project management: Best practices and critical success factors. In *Digital democracy: Concepts, methodologies, tools, and applications* (pp. 1314–1332). Hershey, PA: IGI Global. doi:10.4018/978-1-4666-1740-7.ch065

Akabawi, M. S. (2011). Ghabbour group ERP deployment: Learning from past technology failures. In E. Business Research and Case Center (Ed.), Cases on business and management in the MENA region: New trends and opportunities (pp. 177-203). Hershey, PA: IGI Global. doi:10.4018/978-1-60960-583-4.ch012

Akabawi, M. S. (2013). Ghabbour group ERP deployment: Learning from past technology failures. In *Industrial engineering: Concepts, methodologies, tools, and applications* (pp. 933–958). Hershey, PA: IGI Global. doi:10.4018/978-1-4666-1945-6.ch051

Akbulut, A. Y., & Motwani, J. (2008). Integration and information sharing in e-government. In G. Putnik & M. Cruz-Cunha (Eds.), *Encyclopedia of networked and virtual organizations* (pp. 729–734). Hershey, PA: IGI Global. doi:10.4018/978-1-59904-885-7.ch096

Akers, E. J. (2008). Technology diffusion in public administration. In G. Garson & M. Khosrow-Pour (Eds.), *Handbook of research on public information technology* (pp. 339–348). Hershey, PA: IGI Global. doi:10.4018/978-1-59904-857-4.ch033

Al-Shafi, S. (2008). Free wireless internet park services: An investigation of technology adoption in Qatar from a citizens' perspective. *Journal of Cases on Information Technology, 10*(3), 21–34. doi:10.4018/jcit.2008070103

Al-Shafi, S., & Weerakkody, V. (2009). Implementing free wi-fi in public parks: An empirical study in Qatar. *International Journal of Electronic Government Research, 5*(3), 21–35. doi:10.4018/jegr.2009070102

Aladwani, A. M. (2002). Organizational actions, computer attitudes and end-user satisfaction in public organizations: An empirical study. In C. Snodgrass & E. Szewczak (Eds.), *Human factors in information systems* (pp. 153–168). Hershey, PA: IGI Global. doi:10.4018/978-1-931777-10-0.ch012

Aladwani, A. M. (2002). Organizational actions, computer attitudes, and end-user satisfaction in public organizations: An empirical study. *Journal of Organizational and End User Computing*, *14*(1), 42–49. doi:10.4018/joeuc.2002010104

Allen, B., Juillet, L., Paquet, G., & Roy, J. (2005). E-government and private-public partnerships: Relational challenges and strategic directions. In M. Khosrow-Pour (Ed.), *Practicing e-government: A global perspective* (pp. 364–382). Hershey, PA: IGI Global. doi:10.4018/978-1-59140-637-2.ch016

Alshawaf, A., & Knalil, O. E. (2008). IS success factors and IS organizational impact: Does ownership type matter in Kuwait? *International Journal of Enterprise Information Systems*, *4*(2), 13–33. doi:10.4018/jeis.2008040102

Ambali, A. R. (2009). Digital divide and its implication on Malaysian e-government: Policy initiatives. In H. Rahman (Ed.), *Social and political implications of data mining: Knowledge management in e-government* (pp. 267–287). Hershey, PA: IGI Global. doi:10.4018/978-1-60566-230-5.ch016

Amoretti, F. (2007). Digital international governance. In A. Anttiroiko & M. Malkia (Eds.), *Encyclopedia of digital government* (pp. 365–370). Hershey, PA: IGI Global. doi:10.4018/978-1-59140-789-8.ch056

Amoretti, F. (2008). Digital international governance. In A. Anttiroiko (Ed.), *Electronic government: Concepts, methodologies, tools, and applications* (pp. 688–696). Hershey, PA: IGI Global. doi:10.4018/978-1-59904-947-2.ch058

Amoretti, F. (2008). E-government at supranational level in the European Union. In A. Anttiroiko (Ed.), *Electronic government: Concepts, methodologies, tools, and applications* (pp. 1047–1055). Hershey, PA: IGI Global. doi:10.4018/978-1-59904-947-2.ch079

Amoretti, F. (2008). E-government regimes. In A. Anttiroiko (Ed.), *Electronic government: Concepts, methodologies, tools, and applications* (pp. 3846–3856). Hershey, PA: IGI Global. doi:10.4018/978-1-59904-947-2.ch280

Amoretti, F. (2009). Electronic constitution: A Braudelian perspective. In F. Amoretti (Ed.), *Electronic constitution: Social, cultural, and political implications* (pp. 1–19). Hershey, PA: IGI Global. doi:10.4018/978-1-60566-254-1.ch001

Amoretti, F., & Musella, F. (2009). Institutional isomorphism and new technologies. In M. Khosrow-Pour (Ed.), *Encyclopedia of information science and technology* (2nd ed.; pp. 2066–2071). Hershey, PA: IGI Global. doi:10.4018/978-1-60566-026-4.ch325

Andersen, K. V., & Henriksen, H. Z. (2007). E-government research: Capabilities, interaction, orientation, and values. In D. Norris (Ed.), *Current issues and trends in e-government research* (pp. 269–288). Hershey, PA: IGI Global. doi:10.4018/978-1-59904-283-1.ch013

Anderson, K. V., & Henriksen, H. Z. (2005). The first leg of e-government research: Domains and application areas 1998-2003. *International Journal of Electronic Government Research, 1*(4), 26–44. doi:10.4018/jegr.2005100102

Anttiroiko, A. (2009). Democratic e-governance. In M. Khosrow-Pour (Ed.), *Encyclopedia of information science and technology* (2nd ed., pp. 990–995). Hershey, PA: IGI Global. doi:10.4018/978-1-60566-026-4.ch158

Baker, P. M., Bell, A., & Moon, N. W. (2009). Accessibility issues in municipal wireless networks. In C. Reddick (Ed.), *Handbook of research on strategies for local e-government adoption and implementation: Comparative studies* (pp. 569–588). Hershey, PA: IGI Global. doi:10.4018/978-1-60566-282-4.ch030

Becker, S. A., Keimer, R., & Muth, T. (2010). A case on university and community collaboration: The sci-tech entrepreneurial training services (ETS) program. In S. Becker & R. Niebuhr (Eds.), *Cases on technology innovation: Entrepreneurial successes and pitfalls* (pp. 68–90). Hershey, PA: IGI Global. doi:10.4018/978-1-61520-609-4.ch003

Becker, S. A., Keimer, R., & Muth, T. (2012). A case on university and community collaboration: The sci-tech entrepreneurial training services (ETS) program. In Regional development: Concepts, methodologies, tools, and applications (pp. 947-969). Hershey, PA: IGI Global. doi:10.4018/978-1-4666-0882-5.ch507

Bernardi, R. (2012). Information technology and resistance to public sector reforms: A case study in Kenya. In T. Papadopoulos & P. Kanellis (Eds.), *Public sector reform using information technologies: Transforming policy into practice* (pp. 59–78). Hershey, PA: IGI Global. doi:10.4018/978-1-60960-839-2.ch004

Bernardi, R. (2013). Information technology and resistance to public sector reforms: A case study in Kenya. In *User-driven healthcare: Concepts, methodologies, tools, and applications* (pp. 14–33). Hershey, PA: IGI Global. doi:10.4018/978-1-4666-2770-3.ch002

Bolívar, M. P., Pérez, M. D., & Hernández, A. M. (2012). Municipal e-government services in emerging economies: The Latin-American and Caribbean experiences. In Y. Chen & P. Chu (Eds.), *Electronic governance and cross-boundary collaboration: Innovations and advancing tools* (pp. 198–226). Hershey, PA: IGI Global. doi:10.4018/978-1-60960-753-1.ch011

Borycki, E. M., & Kushniruk, A. W. (2010). Use of clinical simulations to evaluate the impact of health information systems and ubiquitous computing devices upon health professional work. In S. Mohammed & J. Fiaidhi (Eds.), *Ubiquitous health and medical informatics: The ubiquity 2.0 trend and beyond* (pp. 552–573). Hershey, PA: IGI Global. doi:10.4018/978-1-61520-777-0.ch026

Borycki, E. M., & Kushniruk, A. W. (2011). Use of clinical simulations to evaluate the impact of health information systems and ubiquitous computing devices upon health professional work. In *Clinical technologies: Concepts, methodologies, tools and applications* (pp. 532–553). Hershey, PA: IGI Global. doi:10.4018/978-1-60960-561-2.ch220

Buchan, J. (2011). Developing a dynamic and responsive online learning environment: A case study of a large Australian university. In B. Czerkawski (Ed.), *Free and open source software for e-learning: Issues, successes and challenges* (pp. 92–109). Hershey, PA: IGI Global. doi:10.4018/978-1-61520-917-0.ch006

Buenger, A. W. (2008). Digital convergence and cybersecurity policy. In G. Garson & M. Khosrow-Pour (Eds.), *Handbook of research on public information technology* (pp. 395–405). Hershey, PA: IGI Global. doi:10.4018/978-1-59904-857-4.ch038

Burn, J. M., & Loch, K. D. (2002). The societal impact of world wide web - Key challenges for the 21st century. In A. Salehnia (Ed.), *Ethical issues of information systems* (pp. 88–106). Hershey, PA: IGI Global. doi:10.4018/978-1-931777-15-5.ch007

Burn, J. M., & Loch, K. D. (2003). The societal impact of the world wide web-Key challenges for the 21st century. In M. Khosrow-Pour (Ed.), *Advanced topics in information resources management* (Vol. 2, pp. 32–51). Hershey, PA: IGI Global. doi:10.4018/978-1-59140-062-2.ch002

Bwalya, K. J., Du Plessis, T., & Rensleigh, C. (2012). The “quicksilver initiatives” as a framework for e-government strategy design in developing economies. In K. Bwalya & S. Zulu (Eds.), *Handbook of research on e-government in emerging economies: Adoption, e-participation, and legal frameworks* (pp. 605–623). Hershey, PA: IGI Global. doi:10.4018/978-1-4666-0324-0.ch031

Cabotaje, C. E., & Alampay, E. A. (2013). Social media and citizen engagement: Two cases from the Philippines. In S. Saeed & C. Reddick (Eds.), *Human-centered system design for electronic governance* (pp. 225–238). Hershey, PA: IGI Global. doi:10.4018/978-1-4666-3640-8.ch013

Camillo, A., Di Pietro, L., Di Virgilio, F., & Franco, M. (2013). Work-groups conflict at PetroTech-Italy, S.R.L.: The influence of culture on conflict dynamics. In B. Christiansen, E. Turkina, & N. Williams (Eds.), *Cultural and technological influences on global business* (pp. 272–289). Hershey, PA: IGI Global. doi:10.4018/978-1-4666-3966-9.ch015

Capra, E., Francalanci, C., & Marinoni, C. (2008). Soft success factors for m-government. In A. Anttiroiko (Ed.), *Electronic government: Concepts, methodologies, tools, and applications* (pp. 1213–1233). Hershey, PA: IGI Global. doi:10.4018/978-1-59904-947-2.ch089

Cartelli, A. (2009). The implementation of practices with ICT as a new teaching-learning paradigm. In A. Cartelli & M. Palma (Eds.), *Encyclopedia of information communication technology* (pp. 413–417). Hershey, PA: IGI Global. doi:10.4018/978-1-59904-845-1.ch055

Charalabidis, Y., Lampathaki, F., & Askounis, D. (2010). Investigating the landscape in national interoperability frameworks. *International Journal of E-Services and Mobile Applications*, *2*(4), 28–41. doi:10.4018/jesma.2010100103

Charalabidis, Y., Lampathaki, F., & Askounis, D. (2012). Investigating the landscape in national interoperability frameworks. In A. Scupola (Ed.), *Innovative mobile platform developments for electronic services design and delivery* (pp. 218–231). Hershey, PA: IGI Global. doi:10.4018/978-1-4666-1568-7.ch013

Chen, I. (2005). Distance education associations. In C. Howard, J. Boettcher, L. Justice, K. Schenk, P. Rogers, & G. Berg (Eds.), *Encyclopedia of distance learning* (pp. 599–612). Hershey, PA: IGI Global. doi:10.4018/978-1-59140-555-9.ch087

Chen, I. (2008). Distance education associations. In L. Tomei (Ed.), *Online and distance learning: Concepts, methodologies, tools, and applications* (pp. 562–579). Hershey, PA: IGI Global. doi:10.4018/978-1-59904-935-9.ch048

Chen, Y. (2008). Managing IT outsourcing for digital government. In A. Anttiroiko (Ed.), *Electronic government: Concepts, methodologies, tools, and applications* (pp. 3107–3114). Hershey, PA: IGI Global. doi:10.4018/978-1-59904-947-2.ch229

Chen, Y., & Dimitrova, D. V. (2006). Electronic government and online engagement: Citizen interaction with government via web portals. *International Journal of Electronic Government Research, 2*(1), 54–76. doi:10.4018/jegr.2006010104

Chen, Y., & Knepper, R. (2005). Digital government development strategies: Lessons for policy makers from a comparative perspective. In W. Huang, K. Siau, & K. Wei (Eds.), *Electronic government strategies and implementation* (pp. 394–420). Hershey, PA: IGI Global. doi:10.4018/978-1-59140-348-7.ch017

Chen, Y., & Knepper, R. (2008). Digital government development strategies: Lessons for policy makers from a comparative perspective. In H. Rahman (Ed.), *Developing successful ICT strategies: Competitive advantages in a global knowledge-driven society* (pp. 334–356). Hershey, PA: IGI Global. doi:10.4018/978-1-59904-654-9.ch017

Cherian, E. J., & Ryan, T. W. (2014). Incongruent needs: Why differences in the iron-triangle of priorities make health information technology adoption and use difficult. In C. El Morr (Ed.), *Research perspectives on the role of informatics in health policy and management* (pp. 209–221). Hershey, PA: IGI Global. doi:10.4018/978-1-4666-4321-5.ch012

Cho, H. J., & Hwang, S. (2010). Government 2.0 in Korea: Focusing on e-participation services. In C. Reddick (Ed.), *Politics, democracy and e-government: Participation and service delivery* (pp. 94–114). Hershey, PA: IGI Global. doi:10.4018/978-1-61520-933-0.ch006

Chorus, C., & Timmermans, H. (2010). Ubiquitous travel environments and travel control strategies: Prospects and challenges. In M. Wachowicz (Ed.), *Movement-aware applications for sustainable mobility: Technologies and approaches* (pp. 30–51). Hershey, PA: IGI Global. doi:10.4018/978-1-61520-769-5.ch003

Chuanshen, R. (2007). E-government construction and China's administrative litigation act. In A. Anttiroiko & M. Malkia (Eds.), *Encyclopedia of digital government* (pp. 507–510). Hershey, PA: IGI Global. doi:10.4018/978-1-59140-789-8.ch077

Ciaghi, A., & Villafiorita, A. (2012). Law modeling and BPR for public administration improvement. In K. Bwalya & S. Zulu (Eds.), *Handbook of research on e-government in emerging economies: Adoption, e-participation, and legal frameworks* (pp. 391–410). Hershey, PA: IGI Global. doi:10.4018/978-1-4666-0324-0.ch019

Ciaramitaro, B. L., & Skrocki, M. (2012). mHealth: Mobile healthcare. In B. Ciaramitaro (Ed.), Mobile technology consumption: Opportunities and challenges (pp. 99-109). Hershey, PA: IGI Global. doi:10.4018/978-1-61350-150-4.ch007

Comite, U. (2012). Innovative processes and managerial effectiveness of e-procurement in healthcare. In A. Manoharan & M. Holzer (Eds.), *Active citizen participation in e-government: A global perspective* (pp. 206–229). Hershey, PA: IGI Global. doi:10.4018/978-1-4666-0116-1.ch011

Cordella, A. (2013). E-government success: How to account for ICT, administrative rationalization, and institutional change. In J. Gil-Garcia (Ed.), *E-government success factors and measures: Theories, concepts, and methodologies* (pp. 40–51). Hershey, PA: IGI Global. doi:10.4018/978-1-4666-4058-0.ch003

Cropf, R. A. (2009). ICT and e-democracy. In M. Khosrow-Pour (Ed.), *Encyclopedia of information science and technology* (2nd ed.; pp. 1789–1793). Hershey, PA: IGI Global. doi:10.4018/978-1-60566-026-4.ch281

Cropf, R. A. (2009). The virtual public sphere. In M. Pagani (Ed.), *Encyclopedia of multimedia technology and networking* (2nd ed.; pp. 1525–1530). Hershey, PA: IGI Global. doi:10.4018/978-1-60566-014-1.ch206

D'Abundo, M. L. (2013). Electronic health record implementation in the United States healthcare industry: Making the process of change manageable. In V. Wang (Ed.), *Handbook of research on technologies for improving the 21st century workforce: Tools for lifelong learning* (pp. 272–286). Hershey, PA: IGI Global. doi:10.4018/978-1-4666-2181-7.ch018

Damurski, L. (2012). E-participation in urban planning: Online tools for citizen engagement in Poland and in Germany. *International Journal of E-Planning Research*, *1*(3), 40–67. doi:10.4018/ijepr.2012070103

de Almeida, M. O. (2007). E-government strategy in Brazil: Increasing transparency and efficiency through e-government procurement. In M. Gascó-Hernandez (Ed.), *Latin America online: Cases, successes and pitfalls* (pp. 34–82). Hershey, PA: IGI Global. doi:10.4018/978-1-59140-974-8.ch002 de Juana Espinosa, S. (2008). Empirical study of the municipalitites' motivations for adopting online presence. In A. Anttiroiko (Ed.), *Electronic government: Concepts, methodologies, tools, and applications* (pp. 3593–3608). Hershey, PA: IGI Global. doi:10.4018/978-1-59904-947-2.ch262

de Souza Dias, D. (2002). Motivation for using information technology. In C. Snodgrass & E. Szewczak (Eds.), *Human factors in information systems* (pp. 55–60). Hershey, PA: IGI Global. doi:10.4018/978-1-931777-10-0.ch005

Demediuk, P. (2006). Government procurement ICT's impact on the sustainability of SMEs and regional communities. In S. Marshall, W. Taylor, & X. Yu (Eds.), *Encyclopedia of developing regional communities with information and communication technology* (pp. 321–324). Hershey, PA: IGI Global. doi:10.4018/978-1-59140-575-7.ch056

Devonshire, E., Forsyth, H., Reid, S., & Simpson, J. M. (2013). The challenges and opportunities of online postgraduate coursework programs in a traditional university context. In B. Tynan, J. Willems, & R. James (Eds.), *Outlooks and opportunities in blended and distance learning* (pp. 353–368). Hershey, PA: IGI Global. doi:10.4018/978-1-4666-4205-8.ch026

Di Cerbo, F., Scotto, M., Sillitti, A., Succi, G., & Vernazza, T. (2007). Toward a GNU/Linux distribution for corporate environments. In S. Sowe, I. Stamelos, & I. Samoladas (Eds.), *Emerging free and open source software practices* (pp. 215–236). Hershey, PA: IGI Global. doi:10.4018/978-1-59904-210-7.ch010

Diesner, J., & Carley, K. M. (2005). Revealing social structure from texts: Meta-matrix text analysis as a novel method for network text analysis. In V. Narayanan & D. Armstrong (Eds.), *Causal mapping for research in information technology* (pp. 81–108). Hershey, PA: IGI Global. doi:10.4018/978-1-59140-396-8.ch004

Dologite, D. G., Mockler, R. J., Bai, Q., & Viszhanyo, P. F. (2006). IS change agents in practice in a US-Chinese joint venture. In M. Hunter & F. Tan (Eds.), *Advanced topics in global information management* (Vol. 5, pp. 331–352). Hershey, PA: IGI Global. doi:10.4018/978-1-59140-923-6.ch015

Drnevich, P., Brush, T. H., & Luckock, G. T. (2011). Process and structural implications for IT-enabled outsourcing. *International Journal of Strategic Information Technology and Applications*, *2*(4), 30–43. doi:10.4018/jsita.2011100103

Dwivedi, A. N. (2009). *Handbook of research on information technology management and clinical data administration in healthcare* (Vols. 1–2). Hershey, PA: IGI Global. doi:10.4018/978-1-60566-356-2

Elbeltagi, I., McBride, N., & Hardaker, G. (2006). Evaluating the factors affecting DSS usage by senior managers in local authorities in Egypt. In M. Hunter & F. Tan (Eds.), *Advanced topics in global information management* (Vol. 5, pp. 283–307). Hershey, PA: IGI Global. doi:10.4018/978-1-59140-923-6.ch013

Eom, S., & Fountain, J. E. (2013). Enhancing information services through public-private partnerships: Information technology knowledge transfer underlying structures to develop shared services in the U.S. and Korea. In J. Gil-Garcia (Ed.), *E-government success around the world: Cases, empirical studies, and practical recommendations* (pp. 15–40). Hershey, PA: IGI Global. doi:10.4018/978-1-4666-4173-0.ch002

Esteves, T., Leuenberger, D., & Van Leuven, N. (2012). Reaching citizen 2.0: How government uses social media to send public messages during times of calm and times of crisis. In K. Kloby & M. D'Agostino (Eds.), *Citizen 2.0: Public and governmental interaction through web 2.0 technologies* (pp. 250–268). Hershey, PA: IGI Global. doi:10.4018/978-1-4666-0318-9.ch013

Estevez, E., Fillottrani, P., Janowski, T., & Ojo, A. (2012). Government information sharing: A framework for policy formulation. In Y. Chen & P. Chu (Eds.), *Electronic governance and cross-boundary collaboration: Innovations and advancing tools* (pp. 23–55). Hershey, PA: IGI Global. doi:10.4018/978-1-60960-753-1.ch002

Ezz, I. E. (2008). E-governement emerging trends: Organizational challenges. In A. Anttiroiko (Ed.), *Electronic government: Concepts, methodologies, tools, and applications* (pp. 3721–3737). Hershey, PA: IGI Global. doi:10.4018/978-1-59904-947-2.ch269

Fabri, M. (2009). The Italian style of e-justice in a comparative perspective. In A. Martínez & P. Abat (Eds.), *E-justice: Using information communication technologies in the court system* (pp. 1–19). Hershey, PA: IGI Global. doi:10.4018/978-1-59904-998-4.ch001

Fagbe, T., & Adekola, O. D. (2010). Workplace safety and personnel well-being: The impact of information technology. *International Journal of Green Computing*, *1*(1), 28–33. doi:10.4018/jgc.2010010103

Fagbe, T., & Adekola, O. D. (2011). Workplace safety and personnel well-being: The impact of information technology. In *Global business: Concepts, methodologies, tools and applications* (pp. 1438–1444). Hershey, PA: IGI Global. doi:10.4018/978-1-60960-587-2.ch509

Farmer, L. (2008). Affective collaborative instruction with librarians. In S. Kelsey & K. St.Amant (Eds.), *Handbook of research on computer mediated communication* (pp. 15–24). Hershey, PA: IGI Global. doi:10.4018/978-1-59904-863-5.ch002

Favier, L., & Mekhantar, J. (2007). Use of OSS by local e-administration: The French situation. In K. St.Amant & B. Still (Eds.), *Handbook of research on open source software: Technological, economic, and social perspectives* (pp. 428–444). Hershey, PA: IGI Global. doi:10.4018/978-1-59140-999-1.ch033

Fernando, S. (2009). Issues of e-learning in third world countries. In M. Khosrow-Pour (Ed.), *Encyclopedia of information science and technology* (2nd ed.; pp. 2273–2277). Hershey, PA: IGI Global. doi:10.4018/978-1-60566-026-4.ch360

Filho, J. R., & dos Santos, J. R. Junior. (2009). Local e-government in Brazil: Poor interaction and local politics as usual. In C. Reddick (Ed.), *Handbook of research on strategies for local e-government adoption and implementation: Comparative studies* (pp. 863–878). Hershey, PA: IGI Global. doi:10.4018/978-1-60566-282-4.ch045

Fletcher, P. D. (2004). Portals and policy: Implications of electronic access to U.S. federal government information services. In A. Pavlichev & G. Garson (Eds.), *Digital government: Principles and best practices* (pp. 52–62). Hershey, PA: IGI Global. doi:10.4018/978-1-59140-122-3.ch004

Fletcher, P. D. (2008). Portals and policy: Implications of electronic access to U.S. federal government information services. In A. Anttiroiko (Ed.), *Electronic government: Concepts, methodologies, tools, and applications* (pp. 3970–3979). Hershey, PA: IGI Global. doi:10.4018/978-1-59904-947-2.ch289

Forlano, L. (2004). The emergence of digital government: International perspectives. In A. Pavlichev & G. Garson (Eds.), *Digital government: Principles and best practices* (pp. 34–51). Hershey, PA: IGI Global. doi:10.4018/978-1-59140-122-3.ch003

Franzel, J. M., & Coursey, D. H. (2004). Government web portals: Management issues and the approaches of five states. In A. Pavlichev & G. Garson (Eds.), *Digital government: Principles and best practices* (pp. 63–77). Hershey, PA: IGI Global. doi:10.4018/978-1-59140-122-3.ch005

Gaivéo, J. M. (2013). Security of ICTs supporting healthcare activities. In M. Cruz-Cunha, I. Miranda, & P. Gonçalves (Eds.), *Handbook of research on ICTs for human-centered healthcare and social care services* (pp. 208–228). Hershey, PA: IGI Global. doi:10.4018/978-1-4666-3986-7.ch011

Garson, G. D. (1999). *Information technology and computer applications in public administration: Issues and trends*. Hershey, PA: IGI Global. doi:10.4018/978-1-87828-952-0

Garson, G. D. (2003). Toward an information technology research agenda for public administration. In G. Garson (Ed.), *Public information technology: Policy and management issues* (pp. 331–357). Hershey, PA: IGI Global. doi:10.4018/978-1-59140-060-8.ch014

Garson, G. D. (2004). The promise of digital government. In A. Pavlichev & G. Garson (Eds.), *Digital government: Principles and best practices* (pp. 2–15). Hershey, PA: IGI Global. doi:10.4018/978-1-59140-122-3.ch001

Garson, G. D. (2007). An information technology research agenda for public administration. In G. Garson (Ed.), *Modern public information technology systems: Issues and challenges* (pp. 365–392). Hershey, PA: IGI Global. doi:10.4018/978-1-59904-051-6.ch018

Gasco, M. (2007). Civil servants' resistance towards e-government development. In A. Anttiroiko & M. Malkia (Eds.), *Encyclopedia of digital government* (pp. 190–195). Hershey, PA: IGI Global. doi:10.4018/978-1-59140-789-8.ch028

Gasco, M. (2008). Civil servants' resistance towards e-government development. In A. Anttiroiko (Ed.), *Electronic government: Concepts, methodologies, tools, and applications* (pp. 2580–2588). Hershey, PA: IGI Global. doi:10.4018/978-1-59904-947-2.ch190

Ghere, R. K. (2010). Accountability and information technology enactment: Implications for social empowerment. In E. Ferro, Y. Dwivedi, J. Gil-Garcia, & M. Williams (Eds.), *Handbook of research on overcoming digital divides: Constructing an equitable and competitive information society* (pp. 515–532). Hershey, PA: IGI Global. doi:10.4018/978-1-60566-699-0.ch028

Gibson, I. W. (2012). Simulation modeling of healthcare delivery. In A. Kolker & P. Story (Eds.), *Management engineering for effective healthcare delivery: Principles and applications* (pp. 69–89). Hershey, PA: IGI Global. doi:10.4018/978-1-60960-872-9.ch003

Gil-Garcia, J. R. (2007). Exploring e-government benefits and success factors. In A. Anttiroiko & M. Malkia (Eds.), *Encyclopedia of digital government* (pp. 803–811). Hershey, PA: IGI Global. doi:10.4018/978-1-59140-789-8.ch122

Gil-Garcia, J. R., & González Miranda, F. (2010). E-government and opportunities for participation: The case of the Mexican state web portals. In C. Reddick (Ed.), *Politics, democracy and e-government: Participation and service delivery* (pp. 56–74). Hershey, PA: IGI Global. doi:10.4018/978-1-61520-933-0.ch004

Goldfinch, S. (2012). Public trust in government, trust in e-government, and use of e-government. In Z. Yan (Ed.), *Encyclopedia of cyber behavior* (pp. 987–995). Hershey, PA: IGI Global. doi:10.4018/978-1-4666-0315-8.ch081

Goodyear, M. (2012). Organizational change contributions to e-government project transitions. In S. Aikins (Ed.), *Managing e-government projects: Concepts, issues, and best practices* (pp. 1–21). Hershey, PA: IGI Global. doi:10.4018/978-1-4666-0086-7.ch001

Gordon, S., & Mulligan, P. (2003). Strategic models for the delivery of personal financial services: The role of infocracy. In S. Gordon (Ed.), *Computing information technology: The human side* (pp. 220–232). Hershey, PA: IGI Global. doi:10.4018/978-1-93177-752-0.ch014

Gordon, T. F. (2007). Legal knowledge systems. In A. Anttiroiko & M. Malkia (Eds.), *Encyclopedia of digital government* (pp. 1161–1166). Hershey, PA: IGI Global. doi:10.4018/978-1-59140-789-8.ch175

Graham, J. E., & Semich, G. W. (2008). Integrating technology to transform pedagogy: Revisiting the progress of the three phase TUI model for faculty development. In L. Tomei (Ed.), *Adapting information and communication technologies for effective education* (pp. 1–12). Hershey, PA: IGI Global. doi:10.4018/978-1-59904-922-9.ch001

Grandinetti, L., & Pisacane, O. (2012). Web services for healthcare management. In D. Prakash Vidyarthi (Ed.), *Technologies and protocols for the future of internet design: Reinventing the web* (pp. 60–94). Hershey, PA: IGI Global. doi:10.4018/978-1-4666-0203-8.ch004

Groenewegen, P., & Wagenaar, F. P. (2008). VO as an alternative to hierarchy in the Dutch police sector. In G. Putnik & M. Cruz-Cunha (Eds.), *Encyclopedia of networked and virtual organizations* (pp. 1851–1857). Hershey, PA: IGI Global. doi:10.4018/978-1-59904-885-7.ch245

Gronlund, A. (2001). Building an infrastructure to manage electronic services. In S. Dasgupta (Ed.), *Managing internet and intranet technologies in organizations: Challenges and opportunities* (pp. 71–103). Hershey, PA: IGI Global. doi:10.4018/978-1-878289-95-7.ch006

Gronlund, A. (2002). Introduction to electronic government: Design, applications and management. In Å. Grönlund (Ed.), *Electronic government: Design, applications and management* (pp. 1–21). Hershey, PA: IGI Global. doi:10.4018/978-1-930708-19-8.ch001

Gupta, A., Woosley, R., Crk, I., & Sarnikar, S. (2009). An information technology architecture for drug effectiveness reporting and post-marketing surveillance. In J. Tan (Ed.), *Medical informatics: Concepts, methodologies, tools, and applications* (pp. 631–646). Hershey, PA: IGI Global. doi:10.4018/978-1-60566-050-9.ch047

Hallin, A., & Lundevall, K. (2007). mCity: User focused development of mobile services within the city of Stockholm. In I. Kushchu (Ed.), Mobile government: An emerging direction in e-government (pp. 12-29). Hershey, PA: IGI Global. doi:10.4018/978-1-59140-884-0.ch002

Hallin, A., & Lundevall, K. (2009). mCity: User focused development of mobile services within the city of Stockholm. In S. Clarke (Ed.), Evolutionary concepts in end user productivity and performance: Applications for organizational progress (pp. 268-280). Hershey, PA: IGI Global. doi:10.4018/978-1-60566-136-0.ch017

Hallin, A., & Lundevall, K. (2009). mCity: User focused development of mobile services within the city of Stockholm. In D. Taniar (Ed.), Mobile computing: Concepts, methodologies, tools, and applications (pp. 3455-3467). Hershey, PA: IGI Global. doi:10.4018/978-1-60566-054-7.ch253

Hanson, A. (2005). Overcoming barriers in the planning of a virtual library. In M. Khosrow-Pour (Ed.), *Encyclopedia of information science and technology* (pp. 2255–2259). Hershey, PA: IGI Global. doi:10.4018/978-1-59140-553-5.ch397

Haque, A. (2008). Information technology and surveillance: Implications for public administration in a new word order. In T. Loendorf & G. Garson (Eds.), *Patriotic information systems* (pp. 177–185). Hershey, PA: IGI Global. doi:10.4018/978-1-59904-594-8.ch008

Hauck, R. V., Thatcher, S. M., & Weisband, S. P. (2012). Temporal aspects of information technology use: Increasing shift work effectiveness. In J. Wang (Ed.), *Advancing the service sector with evolving technologies: Techniques and principles* (pp. 87–104). Hershey, PA: IGI Global. doi:10.4018/978-1-4666-0044-7.ch006

Hawk, S., & Witt, T. (2006). Telecommunications courses in information systems programs. *International Journal of Information and Communication Technology Education*, *2*(1), 79–92. doi:10.4018/jicte.2006010107

Helms, M. M., Moore, R., & Ahmadi, M. (2009). Information technology (IT) and the healthcare industry: A SWOT analysis. In J. Tan (Ed.), *Medical informatics: Concepts, methodologies, tools, and applications* (pp. 134–152). Hershey, PA: IGI Global. doi:10.4018/978-1-60566-050-9.ch012

Hendrickson, S. M., & Young, M. E. (2014). Electronic records management at a federally funded research and development center. In J. Krueger (Ed.), *Cases on electronic records and resource management implementation in diverse environments* (pp. 334–350). Hershey, PA: IGI Global. doi:10.4018/978-1-4666-4466-3.ch020

Henman, P. (2010). Social policy and information communication technologies. In J. Martin & L. Hawkins (Eds.), *Information communication technologies for human services education and delivery: Concepts and cases* (pp. 215–229). Hershey, PA: IGI Global. doi:10.4018/978-1-60566-735-5.ch014

Hismanoglu, M. (2011). Important issues in online education: E-pedagogy and marketing. In U. Demiray & S. Sever (Eds.), *Marketing online education programs: Frameworks for promotion and communication* (pp. 184–209). Hershey, PA: IGI Global. doi:10.4018/978-1-60960-074-7.ch012

Ho, K. K. (2008). The e-government development, IT strategies, and portals of the Hong Kong SAR government. In A. Anttiroiko (Ed.), *Electronic government: Concepts, methodologies, tools, and applications* (pp. 715–733). Hershey, PA: IGI Global. doi:10.4018/978-1-59904-947-2.ch060

Holden, S. H. (2003). The evolution of information technology management at the federal level: Implications for public administration. In G. Garson (Ed.), *Public information technology: Policy and management issues* (pp. 53–73). Hershey, PA: IGI Global. doi:10.4018/978-1-59140-060-8.ch003

Holden, S. H. (2007). The evolution of federal information technology management literature: Does IT finally matter? In G. Garson (Ed.), *Modern public information technology systems: Issues and challenges* (pp. 17–34). Hershey, PA: IGI Global. doi:10.4018/978-1-59904-051-6.ch002

Holland, J. W. (2009). Automation of American criminal justice. In M. Khosrow-Pour (Ed.), *Encyclopedia of information science and technology* (2nd ed.; pp. 300–302). Hershey, PA: IGI Global. doi:10.4018/978-1-60566-026-4.ch051

Holloway, K. (2013). Fair use, copyright, and academic integrity in an online academic environment. In *Digital rights management: Concepts, methodologies, tools, and applications* (pp. 917–928). Hershey, PA: IGI Global. doi:10.4018/978-1-4666-2136-7.ch044

Horiuchi, C. (2005). E-government databases. In L. Rivero, J. Doorn, & V. Ferraggine (Eds.), *Encyclopedia of database technologies and applications* (pp. 206–210). Hershey, PA: IGI Global. doi:10.4018/978-1-59140-560-3.ch035

Horiuchi, C. (2006). Creating IS quality in government settings. In E. Duggan & J. Reichgelt (Eds.), *Measuring information systems delivery quality* (pp. 311–327). Hershey, PA: IGI Global. doi:10.4018/978-1-59140-857-4.ch014

Hsiao, N., Chu, P., & Lee, C. (2012). Impact of e-governance on businesses: Model development and case study. In *Digital democracy: Concepts, methodologies, tools, and applications* (pp. 1407–1425). Hershey, PA: IGI Global. doi:10.4018/978-1-4666-1740-7.ch070

Huang, T., & Lee, C. (2010). Evaluating the impact of e-government on citizens: Cost-benefit analysis. In C. Reddick (Ed.), *Citizens and e-government: Evaluating policy and management* (pp. 37–52). Hershey, PA: IGI Global. doi:10.4018/978-1-61520-931-6.ch003

Hunter, M. G., Diochon, M., Pugsley, D., & Wright, B. (2002). Unique challenges for small business adoption of information technology: The case of the Nova Scotia ten. In S. Burgess (Ed.), *Managing information technology in small business: Challenges and solutions* (pp. 98–117). Hershey, PA: IGI Global. doi:10.4018/978-1-930708-35-8.ch006

Hurskainen, J. (2003). Integration of business systems and applications in merger and alliance: Case metso automation. In T. Reponen (Ed.), *Information technology enabled global customer service* (pp. 207–225). Hershey, PA: IGI Global. doi:10.4018/978-1-59140-048-6.ch012

Iazzolino, G., & Pietrantonio, R. (2011). The soveria.it project: A best practice of e-government in southern Italy. In D. Piaggesi, K. Sund, & W. Castelnovo (Eds.), *Global strategy and practice of e-governance: Examples from around the world* (pp. 34–56). Hershey, PA: IGI Global. doi:10.4018/978-1-60960-489-9.ch003

Imran, A., & Gregor, S. (2012). A process model for successful e-government adoption in the least developed countries: A case of Bangladesh. In F. Tan (Ed.), *International comparisons of information communication technologies: Advancing applications* (pp. 321–350). Hershey, PA: IGI Global. doi:10.4018/978-1-61350-480-2.ch014

Inoue, Y., & Bell, S. T. (2005). Electronic/digital government innovation, and publishing trends with IT. In M. Khosrow-Pour (Ed.), *Encyclopedia of information science and technology* (pp. 1018–1023). Hershey, PA: IGI Global. doi:10.4018/978-1-59140-553-5.ch180

Islam, M. M., & Ehsan, M. (2013). Understanding e-governance: A theoretical approach. In M. Islam & M. Ehsan (Eds.), *From government to e-governance: Public administration in the digital age* (pp. 38–49). Hershey, PA: IGI Global. doi:10.4018/978-1-4666-1909-8.ch003

Jaeger, B. (2009). E-government and e-democracy in the making. In M. Khosrow-Pour (Ed.), *Encyclopedia of information science and technology* (2nd ed.; pp. 1318–1322). Hershey, PA: IGI Global. doi:10.4018/978-1-60566-026-4.ch208

Jain, R. B. (2007). Revamping the administrative structure and processes in India for online diplomacy. In A. Anttiroiko & M. Malkia (Eds.), *Encyclopedia of digital government* (pp. 1418–1423). Hershey, PA: IGI Global. doi:10.4018/978-1-59140-789-8.ch217

Jain, R. B. (2008). Revamping the administrative structure and processes in India for online diplomacy. In A. Anttiroiko (Ed.), *Electronic government: Concepts, methodologies, tools, and applications* (pp. 3142–3149). Hershey, PA: IGI Global. doi:10.4018/978-1-59904-947-2.ch233

Jauhiainen, J. S., & Inkinen, T. (2009). E-governance and the information society in periphery. In C. Reddick (Ed.), *Handbook of research on strategies for local e-government adoption and implementation: Comparative studies* (pp. 497–514). Hershey, PA: IGI Global. doi:10.4018/978-1-60566-282-4.ch026

Jensen, M. J. (2009). Electronic democracy and citizen influence in government. In C. Reddick (Ed.), *Handbook of research on strategies for local e-government adoption and implementation: Comparative studies* (pp. 288–305). Hershey, PA: IGI Global. doi:10.4018/978-1-60566-282-4.ch015

Jiao, Y., Hurson, A. R., Potok, T. E., & Beckerman, B. G. (2009). Integrating mobile-based systems with healthcare databases. In J. Erickson (Ed.), *Database technologies: Concepts, methodologies, tools, and applications* (pp. 484–504). Hershey, PA: IGI Global. doi:10.4018/978-1-60566-058-5.ch031

Joia, L. A. (2002). A systematic model to integrate information technology into metabusinesses: A case study in the engineering realms. In F. Tan (Ed.), *Advanced topics in global information management* (Vol. 1, pp. 250–267). Hershey, PA: IGI Global. doi:10.4018/978-1-930708-43-3.ch016

Jones, T. H., & Song, I. (2000). Binary equivalents of ternary relationships in entity-relationship modeling: A logical decomposition approach. *Journal of Database Management, 11*(2), 12–19. doi:10.4018/jdm.2000040102

Juana-Espinosa, S. D. (2007). Empirical study of the municipalitites' motivations for adopting online presence. In L. Al-Hakim (Ed.), *Global e-government: Theory, applications and benchmarking* (pp. 261–279). Hershey, PA: IGI Global. doi:10.4018/978-1-59904-027-1.ch015

Jun, K., & Weare, C. (2012). Bridging from e-government practice to e-government research: Past trends and future directions. In K. Bwalya & S. Zulu (Eds.), *Handbook of research on e-government in emerging economies: Adoption, e-participation, and legal frameworks* (pp. 263–289). Hershey, PA: IGI Global. doi:10.4018/978-1-4666-0324-0.ch013

Junqueira, A., Diniz, E. H., & Fernandez, M. (2010). Electronic government implementation projects with multiple agencies: Analysis of the electronic invoice project under PMBOK framework. In J. Cordoba-Pachon & A. Ochoa-Arias (Eds.), *Systems thinking and e-participation: ICT in the governance of society* (pp. 135–153). Hershey, PA: IGI Global. doi:10.4018/978-1-60566-860-4.ch009

Juntunen, A. (2009). Joint service development with the local authorities. In C. Reddick (Ed.), *Handbook of research on strategies for local e-government adoption and implementation: Comparative studies* (pp. 902–920). Hershey, PA: IGI Global. doi:10.4018/978-1-60566-282-4.ch047

Kamel, S. (2001). *Using DSS for crisis management.* Hershey, PA: IGI Global. doi:10.4018/978-1-87828-961-2.ch020

Kamel, S. (2006). DSS for strategic decision making. In M. Khosrow-Pour (Ed.), *Cases on information technology and organizational politics & culture* (pp. 230–246). Hershey, PA: IGI Global. doi:10.4018/978-1-59904-411-8.ch015

Kamel, S. (2009). The software industry in Egypt as a potential contributor to economic growth. In M. Khosrow-Pour (Ed.), *Encyclopedia of information science and technology* (2nd ed.; pp. 3531–3537). Hershey, PA: IGI Global. doi:10.4018/978-1-60566-026-4.ch562

Kamel, S., & Hussein, M. (2008). Xceed: Pioneering the contact center industry in Egypt. *Journal of Cases on Information Technology, 10*(1), 67–91. doi:10.4018/jcit.2008010105

Kamel, S., & Wahba, K. (2003). The use of a hybrid model in web-based education: "The Global campus project. In A. Aggarwal (Ed.), *Web-based education: Learning from experience* (pp. 331–346). Hershey, PA: IGI Global. doi:10.4018/978-1-59140-102-5.ch020

Kardaras, D. K., & Papathanassiou, E. A. (2008). An exploratory study of the e-government services in Greece. In G. Garson & M. Khosrow-Pour (Eds.), *Handbook of research on public information technology* (pp. 162–174). Hershey, PA: IGI Global. doi:10.4018/978-1-59904-857-4.ch016

Kassahun, A. E., Molla, A., & Sarkar, P. (2012). Government process reengineering: What we know and what we need to know. In *Digital democracy: Concepts, methodologies, tools, and applications* (pp. 1730–1752). Hershey, PA: IGI Global. doi:10.4018/978-1-4666-1740-7.ch086

Khan, B. (2005). Technological issues. In B. Khan (Ed.), *Managing e-learning strategies: Design, delivery, implementation and evaluation* (pp. 154–180). Hershey, PA: IGI Global. doi:10.4018/978-1-59140-634-1.ch004

Khasawneh, A., Bsoul, M., Obeidat, I., & Al Azzam, I. (2012). Technology fears: A study of e-commerce loyalty perception by Jordanian customers. In J. Wang (Ed.), *Advancing the service sector with evolving technologies: Techniques and principles* (pp. 158–165). Hershey, PA: IGI Global. doi:10.4018/978-1-4666-0044-7.ch010

Khatibi, V., & Montazer, G. A. (2012). E-research methodology. In A. Juan, T. Daradoumis, M. Roca, S. Grasman, & J. Faulin (Eds.), *Collaborative and distributed e-research: Innovations in technologies, strategies and applications* (pp. 62–81). Hershey, PA: IGI Global. doi:10.4018/978-1-4666-0125-3.ch003

Kidd, T. (2011). The dragon in the school's backyard: A review of literature on the uses of technology in urban schools. In L. Tomei (Ed.), *Online courses and ICT in education: Emerging practices and applications* (pp. 242–257). Hershey, PA: IGI Global. doi:10.4018/978-1-60960-150-8.ch019

Kidd, T. T. (2010). My experience tells the story: Exploring technology adoption from a qualitative perspective - A pilot study. In H. Song & T. Kidd (Eds.), *Handbook of research on human performance and instructional technology* (pp. 247–262). Hershey, PA: IGI Global. doi:10.4018/978-1-60566-782-9.ch015

Kieley, B., Lane, G., Paquet, G., & Roy, J. (2002). e-Government in Canada: Services online or public service renewal? In Å. Grönlund (Ed.), Electronic government: Design, applications and management (pp. 340-355). Hershey, PA: IGI Global. doi:10.4018/978-1-930708-19-8.ch016

Kim, P. (2012). "Stay out of the way! My kid is video blogging through a phone!": A lesson learned from math tutoring social media for children in underserved communities. In *Wireless technologies: Concepts, methodologies, tools and applications* (pp. 1415–1428). Hershey, PA: IGI Global. doi:10.4018/978-1-61350-101-6.ch517

Kirlidog, M. (2010). Financial aspects of national ICT strategies. In S. Kamel (Ed.), *E-strategies for technological diffusion and adoption: National ICT approaches for socioeconomic development* (pp. 277–292). Hershey, PA: IGI Global. doi:10.4018/978-1-60566-388-3.ch016

Kisielnicki, J. (2006). Transfer of information and knowledge in the project management. In E. Coakes & S. Clarke (Eds.), *Encyclopedia of communities of practice in information and knowledge management* (pp. 544–551). Hershey, PA: IGI Global. doi:10.4018/978-1-59140-556-6.ch091

Kittner, M., & Van Slyke, C. (2006). Reorganizing information technology services in an academic environment. In M. Khosrow-Pour (Ed.), *Cases on the human side of information technology* (pp. 49–66). Hershey, PA: IGI Global. doi:10.4018/978-1-59904-405-7.ch004

Knoell, H. D. (2008). Semi virtual workplaces in German financial service enterprises. In P. Zemliansky & K. St.Amant (Eds.), *Handbook of research on virtual workplaces and the new nature of business practices* (pp. 570–581). Hershey, PA: IGI Global. doi:10.4018/978-1-59904-893-2.ch041

Koh, S. L., & Maguire, S. (2009). Competing in the age of information technology in a developing economy: Experiences of an Indian bank. In S. Koh & S. Maguire (Eds.), *Information and communication technologies management in turbulent business environments* (pp. 326–350). Hershey, PA: IGI Global. doi:10.4018/978-1-60566-424-8.ch018

Kollmann, T., & Häsel, M. (2009). Competence of information technology professionals in internet-based ventures. In I. Lee (Ed.), *Electronic business: Concepts, methodologies, tools, and applications* (pp. 1905–1919). Hershey, PA: IGI Global. doi:10.4018/978-1-60566-056-1.ch118

Kollmann, T., & Häsel, M. (2009). Competence of information technology professionals in internet-based ventures. In A. Cater-Steel (Ed.), *Information technology governance and service management: Frameworks and adaptations* (pp. 239–253). Hershey, PA: IGI Global. doi:10.4018/978-1-60566-008-0.ch013

Kollmann, T., & Häsel, M. (2010). Competence of information technology professionals in internet-based ventures. In *Electronic services: Concepts, methodologies, tools and applications* (pp. 1551–1565). Hershey, PA: IGI Global. doi:10.4018/978-1-61520-967-5.ch094

Kraemer, K., & King, J. L. (2006). Information technology and administrative reform: Will e-government be different? *International Journal of Electronic Government Research, 2*(1), 1–20. doi:10.4018/jegr.2006010101

Kraemer, K., & King, J. L. (2008). Information technology and administrative reform: Will e-government be different? In D. Norris (Ed.), *E-government research: Policy and management* (pp. 1–20). Hershey, PA: IGI Global. doi:10.4018/978-1-59904-913-7.ch001

Lampathaki, F., Tsiakaliaris, C., Stasis, A., & Charalabidis, Y. (2011). National interoperability frameworks: The way forward. In Y. Charalabidis (Ed.), *Interoperability in digital public services and administration: Bridging e-government and e-business* (pp. 1–24). Hershey, PA: IGI Global. doi:10.4018/978-1-61520-887-6.ch001

Lan, Z., & Scott, C. R. (1996). The relative importance of computer-mediated information versus conventional non-computer-mediated information in public managerial decision making. *Information Resources Management Journal*, *9*(1), 27–0. doi:10.4018/irmj.1996010103

Law, W. (2004). *Public sector data management in a developing economy*. Hershey, PA: IGI Global. doi:10.4018/978-1-59140-259-6.ch034

Law, W. K. (2005). Information resources development challenges in a cross-cultural environment. In M. Khosrow-Pour (Ed.), *Encyclopedia of information science and technology* (pp. 1476–1481). Hershey, PA: IGI Global. doi:10.4018/978-1-59140-553-5.ch259

Law, W. K. (2009). Cross-cultural challenges for information resources management. In M. Khosrow-Pour (Ed.), *Encyclopedia of information science and technology* (2nd ed.; pp. 840–846). Hershey, PA: IGI Global. doi:10.4018/978-1-60566-026-4.ch136

Law, W. K. (2011). Cross-cultural challenges for information resources management. In *Global business: Concepts, methodologies, tools and applications* (pp. 1924–1932). Hershey, PA: IGI Global. doi:10.4018/978-1-60960-587-2.ch704

Malkia, M., & Savolainen, R. (2004). eTransformation in government, politics and society: Conceptual framework and introduction. In M. Malkia, A. Anttiroiko, & R. Savolainen (Eds.), eTransformation in governance: New directions in government and politics (pp. 1-21). Hershey, PA: IGI Global. doi:10.4018/978-1-59140-130-8.ch001

Mandujano, S. (2011). Network manageability security. In D. Kar & M. Syed (Eds.), *Network security, administration and management: Advancing technology and practice* (pp. 158–181). Hershey, PA: IGI Global. doi:10.4018/978-1-60960-777-7.ch009

Marich, M. J., Schooley, B. L., & Horan, T. A. (2012). A normative enterprise architecture for guiding end-to-end emergency response decision support. In M. Jennex (Ed.), *Managing crises and disasters with emerging technologies: Advancements* (pp. 71–87). Hershey, PA: IGI Global. doi:10.4018/978-1-4666-0167-3.ch006

Markov, R., & Okujava, S. (2008). Costs, benefits, and risks of e-government portals. In G. Putnik & M. Cruz-Cunha (Eds.), *Encyclopedia of networked and virtual organizations* (pp. 354–363). Hershey, PA: IGI Global. doi:10.4018/978-1-59904-885-7.ch047

Martin, N., & Rice, J. (2013). Evaluating and designing electronic government for the future: Observations and insights from Australia. In V. Weerakkody (Ed.), *E-government services design, adoption, and evaluation* (pp. 238–258). Hershey, PA: IGI Global. doi:10.4018/978-1-4666-2458-0.ch014

i. Martinez, A. C. (2008). Accessing administration's information via internet in Spain. In F. Tan (Ed.), *Global information technologies: Concepts, methodologies, tools, and applications* (pp. 2558–2573). Hershey, PA: IGI Global. doi:10.4018/978-1-59904-939-7.ch186

Mbarika, V. W., Meso, P. N., & Musa, P. F. (2006). A disconnect in stakeholders' perceptions from emerging realities of teledensity growth in Africa's least developed countries. In M. Hunter & F. Tan (Eds.), *Advanced topics in global information management* (Vol. 5, pp. 263–282). Hershey, PA: IGI Global. doi:10.4018/978-1-59140-923-6.ch012

Mbarika, V. W., Meso, P. N., & Musa, P. F. (2008). A disconnect in stakeholders' perceptions from emerging realities of teledensity growth in Africa's least developed countries. In F. Tan (Ed.), *Global information technologies: Concepts, methodologies, tools, and applications* (pp. 2948–2962). Hershey, PA: IGI Global. doi:10.4018/978-1-59904-939-7.ch209

Means, T., Olson, E., & Spooner, J. (2013). Discovering ways that don't work on the road to success: Strengths and weaknesses revealed by an active learning studio classroom project. In A. Benson, J. Moore, & S. Williams van Rooij (Eds.), *Cases on educational technology planning, design, and implementation: A project management perspective* (pp. 94–113). Hershey, PA: IGI Global. doi:10.4018/978-1-4666-4237-9.ch006

Melitski, J., Holzer, M., Kim, S., Kim, C., & Rho, S. (2008). Digital government worldwide: An e-government assessment of municipal web sites. In G. Garson & M. Khosrow-Pour (Eds.), *Handbook of research on public information technology* (pp. 790–804). Hershey, PA: IGI Global. doi:10.4018/978-1-59904-857-4.ch069

Memmola, M., Palumbo, G., & Rossini, M. (2009). Web & RFID technology: New frontiers in costing and process management for rehabilitation medicine. In L. Al-Hakim & M. Memmola (Eds.), *Business web strategy: Design, alignment, and application* (pp. 145–169). Hershey, PA: IGI Global. doi:10.4018/978-1-60566-024-0.ch008

Meng, Z., Fahong, Z., & Lei, L. (2008). Information technology and environment. In Y. Kurihara, S. Takaya, H. Harui, & H. Kamae (Eds.), *Information technology and economic development* (pp. 201–212). Hershey, PA: IGI Global. doi:10.4018/978-1-59904-579-5.ch014

Mentzingen de Moraes, A. J., Ferneda, E., Costa, I., & Spinola, M. D. (2011). Practical approach for implementation of governance process in IT: Information technology areas. In N. Shi & G. Silvius (Eds.), *Enterprise IT governance, business value and performance measurement* (pp. 19–40). Hershey, PA: IGI Global. doi:10.4018/978-1-60566-346-3.ch002

Merwin, G. A. Jr, McDonald, J. S., & Odera, L. C. (2008). Economic development: Government's cutting edge in IT. In M. Raisinghani (Ed.), *Handbook of research on global information technology management in the digital economy* (pp. 1–37). Hershey, PA: IGI Global. doi:10.4018/978-1-59904-875-8.ch001

Meso, P., & Duncan, N. (2002). Can national information infrastructures enhance social development in the least developed countries? An empirical investigation. In M. Dadashzadeh (Ed.), *Information technology management in developing countries* (pp. 23–51). Hershey, PA: IGI Global. doi:10.4018/978-1-931777-03-2.ch002

Meso, P. N., & Duncan, N. B. (2002). Can national information infrastructures enhance social development in the least developed countries? In F. Tan (Ed.), *Advanced topics in global information management* (Vol. 1, pp. 207–226). Hershey, PA: IGI Global. doi:10.4018/978-1-930708-43-3.ch014

Middleton, M. (2008). Evaluation of e-government web sites. In G. Garson & M. Khosrow-Pour (Eds.), *Handbook of research on public information technology* (pp. 699–710). Hershey, PA: IGI Global. doi:10.4018/978-1-59904-857-4.ch063

Mingers, J. (2010). Pluralism, realism, and truth: The keys to knowledge in information systems research. In D. Paradice (Ed.), *Emerging systems approaches in information technologies: Concepts, theories, and applications* (pp. 86–98). Hershey, PA: IGI Global. doi:10.4018/978-1-60566-976-2.ch006

Mital, K. M. (2012). ICT, unique identity and inclusive growth: An Indian perspective. In A. Manoharan & M. Holzer (Eds.), *E-governance and civic engagement: Factors and determinants of e-democracy* (pp. 584–612). Hershey, PA: IGI Global. doi:10.4018/978-1-61350-083-5.ch029

Mizell, A. P. (2008). Helping close the digital divide for financially disadvantaged seniors. In F. Tan (Ed.), *Global information technologies: Concepts, methodologies, tools, and applications* (pp. 2396–2402). Hershey, PA: IGI Global. doi:10.4018/978-1-59904-939-7.ch173

Molinari, F., Wills, C., Koumpis, A., & Moumtzi, V. (2011). A citizen-centric platform to support networking in the area of e-democracy. In H. Rahman (Ed.), *Cases on adoption, diffusion and evaluation of global e-governance systems: Impact at the grass roots* (pp. 282–302). Hershey, PA: IGI Global. doi:10.4018/978-1-61692-814-8.ch014

Molinari, F., Wills, C., Koumpis, A., & Moumtzi, V. (2013). A citizen-centric platform to support networking in the area of e-democracy. In H. Rahman (Ed.), *Cases on progressions and challenges in ICT utilization for citizen-centric governance* (pp. 265–297). Hershey, PA: IGI Global. doi:10.4018/978-1-4666-2071-1.ch013

Monteverde, F. (2010). The process of e-government public policy inclusion in the governmental agenda: A framework for assessment and case study. In J. Cordoba-Pachon & A. Ochoa-Arias (Eds.), *Systems thinking and e-participation: ICT in the governance of society* (pp. 233–245). Hershey, PA: IGI Global. doi:10.4018/978-1-60566-860-4.ch015

Moodley, S. (2008). Deconstructing the South African government's ICT for development discourse. In A. Anttiroiko (Ed.), *Electronic government: Concepts, methodologies, tools, and applications* (pp. 622–631). Hershey, PA: IGI Global. doi:10.4018/978-1-59904-947-2.ch053

Moodley, S. (2008). Deconstructing the South African government's ICT for development discourse. In C. Van Slyke (Ed.), *Information communication technologies: Concepts, methodologies, tools, and applications* (pp. 816–825). Hershey, PA: IGI Global. doi:10.4018/978-1-59904-949-6.ch052

Mora, M., Cervantes-Perez, F., Gelman-Muravchik, O., Forgionne, G. A., & Mejia-Olvera, M. (2003). DMSS implementation research: A conceptual analysis of the contributions and limitations of the factor-based and stage-based streams. In G. Forgionne, J. Gupta, & M. Mora (Eds.), *Decision-making support systems: Achievements and challenges for the new decade* (pp. 331–356). Hershey, PA: IGI Global. doi:10.4018/978-1-59140-045-5.ch020

Mörtberg, C., & Elovaara, P. (2010). Attaching people and technology: Between e and government. In S. Booth, S. Goodman, & G. Kirkup (Eds.), *Gender issues in learning and working with information technology: Social constructs and cultural contexts* (pp. 83–98). Hershey, PA: IGI Global. doi:10.4018/978-1-61520-813-5.ch005

Murphy, J., Harper, E., Devine, E. C., Burke, L. J., & Hook, M. L. (2011). Case study: Lessons learned when embedding evidence-based knowledge in a nurse care planning and documentation system. In A. Cashin & R. Cook (Eds.), *Evidence-based practice in nursing informatics: Concepts and applications* (pp. 174–190). Hershey, PA: IGI Global. doi:10.4018/978-1-60960-034-1.ch014

Mutula, S. M. (2013). E-government's role in poverty alleviation: Case study of South Africa. In H. Rahman (Ed.), *Cases on progressions and challenges in ICT utilization for citizen-centric governance* (pp. 44–68). Hershey, PA: IGI Global. doi:10.4018/978-1-4666-2071-1.ch003

Nath, R., & Angeles, R. (2005). Relationships between supply characteristics and buyer-supplier coupling in e-procurement: An empirical analysis. *International Journal of E-Business Research, 1*(2), 40–55. doi:10.4018/jebr.2005040103

Nissen, M. E. (2006). Application cases in government. In M. Nissen (Ed.), *Harnessing knowledge dynamics: Principled organizational knowing & learning* (pp. 152–181). Hershey, PA: IGI Global. doi:10.4018/978-1-59140-773-7.ch008

Norris, D. F. (2003). Leading-edge information technologies and American local governments. In G. Garson (Ed.), *Public information technology: Policy and management issues* (pp. 139–169). Hershey, PA: IGI Global. doi:10.4018/978-1-59140-060-8.ch007

Norris, D. F. (2008). Information technology among U.S. local governments. In G. Garson & M. Khosrow-Pour (Eds.), *Handbook of research on public information technology* (pp. 132–144). Hershey, PA: IGI Global. doi:10.4018/978-1-59904-857-4.ch013

Northrop, A. (1999). The challenge of teaching information technology in public administration graduate programs. In G. Garson (Ed.), *Information technology and computer applications in public administration: Issues and trends* (pp. 1–22). Hershey, PA: IGI Global. doi:10.4018/978-1-87828-952-0.ch001

Northrop, A. (2003). Information technology and public administration: The view from the profession. In G. Garson (Ed.), *Public information technology: Policy and management issues* (pp. 1–19). Hershey, PA: IGI Global. doi:10.4018/978-1-59140-060-8.ch001

Northrop, A. (2007). Lip service? How PA journals and textbooks view information technology. In G. Garson (Ed.), *Modern public information technology systems: Issues and challenges* (pp. 1–16). Hershey, PA: IGI Global. doi:10.4018/978-1-59904-051-6.ch001

Null, E. (2013). Legal and political barriers to municipal networks in the United States. In A. Abdelaal (Ed.), *Social and economic effects of community wireless networks and infrastructures* (pp. 27–56). Hershey, PA: IGI Global. doi:10.4018/978-1-4666-2997-4.ch003

Okunoye, A., Frolick, M., & Crable, E. (2006). ERP implementation in higher education: An account of pre-implementation and implementation phases. *Journal of Cases on Information Technology*, *8*(2), 110–132. doi:10.4018/jcit.2006040106

Olasina, G. (2012). A review of egovernment services in Nigeria. In A. Tella & A. Issa (Eds.), *Library and information science in developing countries: Contemporary issues* (pp. 205–221). Hershey, PA: IGI Global. doi:10.4018/978-1-61350-335-5.ch015

Orgeron, C. P. (2008). A model for reengineering IT job classes in state government. In G. Garson & M. Khosrow-Pour (Eds.), *Handbook of research on public information technology* (pp. 735–746). Hershey, PA: IGI Global. doi:10.4018/978-1-59904-857-4.ch066

Owsinski, J. W., & Pielak, A. M. (2011). Local authority websites in rural areas: Measuring quality and functionality, and assessing the role. In Z. Andreopoulou, B. Manos, N. Polman, & D. Viaggi (Eds.), *Agricultural and environmental informatics, governance and management: Emerging research applications* (pp. 39–60). Hershey, PA: IGI Global. doi:10.4018/978-1-60960-621-3.ch003

Owsiński, J. W., Pielak, A. M., Sęp, K., & Stańczak, J. (2014). Local web-based networks in rural municipalities: Extension, density, and meaning. In Z. Andreopoulou, V. Samathrakis, S. Louca, & M. Vlachopoulou (Eds.), *E-innovation for sustainable development of rural resources during global economic crisis* (pp. 126–151). Hershey, PA: IGI Global. doi:10.4018/978-1-4666-4550-9.ch011

Pagani, M., & Pasinetti, C. (2008). Technical and functional quality in the development of t-government services. In A. Anttiroiko (Ed.), *Electronic government: Concepts, methodologies, tools, and applications* (pp. 2943–2965). Hershey, PA: IGI Global. doi:10.4018/978-1-59904-947-2.ch220

Pani, A. K., & Agrahari, A. (2005). On e-markets in emerging economy: An Indian experience. In M. Khosrow-Pour (Ed.), *Advanced topics in electronic commerce* (Vol. 1, pp. 287–299). Hershey, PA: IGI Global. doi:10.4018/978-1-59140-819-2.ch015

Papadopoulos, T., Angelopoulos, S., & Kitsios, F. (2011). A strategic approach to e-health interoperability using e-government frameworks. In A. Lazakidou, K. Siassiakos, & K. Ioannou (Eds.), *Wireless technologies for ambient assisted living and healthcare: Systems and applications* (pp. 213–229). Hershey, PA: IGI Global. doi:10.4018/978-1-61520-805-0.ch012

Papadopoulos, T., Angelopoulos, S., & Kitsios, F. (2013). A strategic approach to e-health interoperability using e-government frameworks. In *User-driven healthcare: Concepts, methodologies, tools, and applications* (pp. 791–807). Hershey, PA: IGI Global. doi:10.4018/978-1-4666-2770-3.ch039

Papaleo, G., Chiarella, D., Aiello, M., & Caviglione, L. (2012). Analysis, development and deployment of statistical anomaly detection techniques for real e-mail traffic. In T. Chou (Ed.), *Information assurance and security technologies for risk assessment and threat management: Advances* (pp. 47–71). Hershey, PA: IGI Global. doi:10.4018/978-1-61350-507-6.ch003

Papp, R. (2003). Information technology & FDA compliance in the pharmaceutical industry. In M. Khosrow-Pour (Ed.), *Annals of cases on information technology* (Vol. 5, pp. 262–273). Hershey, PA: IGI Global. doi:10.4018/978-1-59140-061-5.ch017

Parsons, T. W. (2007). Developing a knowledge management portal. In A. Tatnall (Ed.), *Encyclopedia of portal technologies and applications* (pp. 223–227). Hershey, PA: IGI Global. doi:10.4018/978-1-59140-989-2.ch039

Passaris, C. E. (2007). Immigration and digital government. In A. Anttiroiko & M. Malkia (Eds.), *Encyclopedia of digital government* (pp. 988–994). Hershey, PA: IGI Global. doi:10.4018/978-1-59140-789-8.ch148

Pavlichev, A. (2004). The e-government challenge for public administration. In A. Pavlichev & G. Garson (Eds.), *Digital government: Principles and best practices* (pp. 276–290). Hershey, PA: IGI Global. doi:10.4018/978-1-59140-122-3.ch018

Penrod, J. I., & Harbor, A. F. (2000). Designing and implementing a learning organization-oriented information technology planning and management process. In L. Petrides (Ed.), *Case studies on information technology in higher education: Implications for policy and practice* (pp. 7–19). Hershey, PA: IGI Global. doi:10.4018/978-1-878289-74-2.ch001

Planas-Silva, M. D., & Joseph, R. C. (2011). Perspectives on the adoption of electronic resources for use in clinical trials. In M. Guah (Ed.), *Healthcare delivery reform and new technologies: Organizational initiatives* (pp. 19–28). Hershey, PA: IGI Global. doi:10.4018/978-1-60960-183-6.ch002

Pomazalová, N., & Rejman, S. (2013). The rationale behind implementation of new electronic tools for electronic public procurement. In N. Pomazalová (Ed.), *Public sector transformation processes and internet public procurement: Decision support systems* (pp. 85–117). Hershey, PA: IGI Global. doi:10.4018/978-1-4666-2665-2.ch006

Postorino, M. N. (2012). City competitiveness and airport: Information science perspective. In M. Bulu (Ed.), *City competitiveness and improving urban subsystems: Technologies and applications* (pp. 61–83). Hershey, PA: IGI Global. doi:10.4018/978-1-61350-174-0.ch004

Poupa, C. (2002). Electronic government in Switzerland: Priorities for 2001-2005 - Electronic voting and federal portal. In Å. Grönlund (Ed.), *Electronic government: Design, applications and management* (pp. 356–369). Hershey, PA: IGI Global. doi:10.4018/978-1-930708-19-8.ch017

Powell, S. R. (2010). Interdisciplinarity in telecommunications and networking. In *Networking and telecommunications: Concepts, methodologies, tools and applications* (pp. 33–40). Hershey, PA: IGI Global. doi:10.4018/978-1-60566-986-1.ch004

Priya, P. S., & Mathiyalagan, N. (2011). A study of the implementation status of two e-governance projects in land revenue administration in India. In M. Shareef, V. Kumar, U. Kumar, & Y. Dwivedi (Eds.), *Stakeholder adoption of e-government services: Driving and resisting factors* (pp. 214–230). Hershey, PA: IGI Global. doi:10.4018/978-1-60960-601-5.ch011

Prysby, C., & Prysby, N. (2000). Electronic mail, employee privacy and the workplace. In L. Janczewski (Ed.), *Internet and intranet security management: Risks and solutions* (pp. 251–270). Hershey, PA: IGI Global. doi:10.4018/978-1-878289-71-1.ch009

Prysby, C. L., & Prysby, N. D. (2003). Electronic mail in the public workplace: Issues of privacy and public disclosure. In G. Garson (Ed.), *Public information technology: Policy and management issues* (pp. 271–298). Hershey, PA: IGI Global. doi:10.4018/978-1-59140-060-8.ch012

Prysby, C. L., & Prysby, N. D. (2007). You have mail, but who is reading it? Issues of e-mail in the public workplace. In G. Garson (Ed.), *Modern public information technology systems: Issues and challenges* (pp. 312–336). Hershey, PA: IGI Global. doi:10.4018/978-1-59904-051-6.ch016

Radl, A., & Chen, Y. (2005). Computer security in electronic government: A state-local education information system. *International Journal of Electronic Government Research, 1*(1), 79–99. doi:10.4018/jegr.2005010105

Rahman, H. (2008). Information dynamics in developing countries. In C. Van Slyke (Ed.), *Information communication technologies: Concepts, methodologies, tools, and applications* (pp. 104–114). Hershey, PA: IGI Global. doi:10.4018/978-1-59904-949-6.ch008

Ramanathan, J. (2009). Adaptive IT architecture as a catalyst for network capability in government. In P. Saha (Ed.), *Advances in government enterprise architecture* (pp. 149–172). Hershey, PA: IGI Global. doi:10.4018/978-1-60566-068-4.ch007

Ramos, I., & Berry, D. M. (2006). Social construction of information technology supporting work. In M. Khosrow-Pour (Ed.), *Cases on information technology: Lessons learned* (Vol. 7, pp. 36–52). Hershey, PA: IGI Global. doi:10.4018/978-1-59140-673-0.ch003

Ray, D., Gulla, U., Gupta, M. P., & Dash, S. S. (2009). Interoperability and constituents of interoperable systems in public sector. In V. Weerakkody, M. Janssen, & Y. Dwivedi (Eds.), *Handbook of research on ICT-enabled transformational government: A global perspective* (pp. 175–195). Hershey, PA: IGI Global. doi:10.4018/978-1-60566-390-6.ch010

Reddick, C. G. (2007). E-government and creating a citizen-centric government: A study of federal government CIOs. In G. Garson (Ed.), *Modern public information technology systems: Issues and challenges* (pp. 143–165). Hershey, PA: IGI Global. doi:10.4018/978-1-59904-051-6.ch008

Reddick, C. G. (2010). Citizen-centric e-government. In C. Reddick (Ed.), *Homeland security preparedness and information systems: Strategies for managing public policy* (pp. 45–75). Hershey, PA: IGI Global. doi:10.4018/978-1-60566-834-5.ch002

Reddick, C. G. (2010). E-government and creating a citizen-centric government: A study of federal government CIOs. In C. Reddick (Ed.), *Homeland security preparedness and information systems: Strategies for managing public policy* (pp. 230–250). Hershey, PA: IGI Global. doi:10.4018/978-1-60566-834-5.ch012

Reddick, C. G. (2010). Perceived effectiveness of e-government and its usage in city governments: Survey evidence from information technology directors. In C. Reddick (Ed.), *Homeland security preparedness and information systems: Strategies for managing public policy* (pp. 213–229). Hershey, PA: IGI Global. doi:10.4018/978-1-60566-834-5.ch011

Reddick, C. G. (2012). Customer relationship management adoption in local governments in the United States. In S. Chhabra & M. Kumar (Eds.), *Strategic enterprise resource planning models for e-government: Applications and methodologies* (pp. 111–124). Hershey, PA: IGI Global. doi:10.4018/978-1-60960-863-7.ch008

Reeder, F. S., & Pandy, S. M. (2008). Identifying effective funding models for e-government. In A. Anttiroiko (Ed.), *Electronic government: Concepts, methodologies, tools, and applications* (pp. 1108–1138). Hershey, PA: IGI Global. doi:10.4018/978-1-59904-947-2.ch083

Riesco, D., Acosta, E., & Montejano, G. (2003). An extension to a UML activity graph from workflow. In L. Favre (Ed.), *UML and the unified process* (pp. 294–314). Hershey, PA: IGI Global. doi:10.4018/978-1-93177-744-5.ch015

Ritzhaupt, A. D., & Gill, T. G. (2008). A hybrid and novel approach to teaching computer programming in MIS curriculum. In S. Negash, M. Whitman, A. Woszczynski, K. Hoganson, & H. Mattord (Eds.), *Handbook of distance learning for real-time and asynchronous information technology education* (pp. 259–281). Hershey, PA: IGI Global. doi:10.4018/978-1-59904-964-9.ch014

Roche, E. M. (1993). International computing and the international regime. *Journal of Global Information Management, 1*(2), 33–44. doi:10.4018/jgim.1993040103

Rocheleau, B. (2007). Politics, accountability, and information management. In G. Garson (Ed.), *Modern public information technology systems: Issues and challenges* (pp. 35–71). Hershey, PA: IGI Global. doi:10.4018/978-1-59904-051-6.ch003

Rodrigues Filho, J. (2010). E-government in Brazil: Reinforcing dominant institutions or reducing citizenship? In C. Reddick (Ed.), *Politics, democracy and e-government: Participation and service delivery* (pp. 347–362). Hershey, PA: IGI Global. doi:10.4018/978-1-61520-933-0.ch021

Rodriguez, S. R., & Thorp, D. A. (2013). eLearning for industry: A case study of the project management process. In A. Benson, J. Moore, & S. Williams van Rooij (Eds.), Cases on educational technology planning, design, and implementation: A project management perspective (pp. 319-342). Hershey, PA: IGI Global. doi:10.4018/978-1-4666-4237-9.ch017

Roman, A. V. (2013). Delineating three dimensions of e-government success: Security, functionality, and transformation. In J. Gil-Garcia (Ed.), *E-government success factors and measures: Theories, concepts, and methodologies* (pp. 171–192). Hershey, PA: IGI Global. doi:10.4018/978-1-4666-4058-0.ch010

Ross, S. C., Tyran, C. K., & Auer, D. J. (2008). Up in smoke: Rebuilding after an IT disaster. In H. Nemati (Ed.), *Information security and ethics: Concepts, methodologies, tools, and applications* (pp. 3659–3675). Hershey, PA: IGI Global. doi:10.4018/978-1-59904-937-3.ch248

Ross, S. C., Tyran, C. K., Auer, D. J., Junell, J. M., & Williams, T. G. (2005). Up in smoke: Rebuilding after an IT disaster. *Journal of Cases on Information Technology, 7*(2), 31–49. doi:10.4018/jcit.2005040103

Roy, J. (2008). Security, sovereignty, and continental interoperability: Canada's elusive balance. In T. Loendorf & G. Garson (Eds.), *Patriotic information systems* (pp. 153–176). Hershey, PA: IGI Global. doi:10.4018/978-1-59904-594-8.ch007

Rubeck, R. F., & Miller, G. A. (2009). vGOV: Remote video access to government services. In A. Scupola (Ed.), Cases on managing e-services (pp. 253-268). Hershey, PA: IGI Global. doi:10.4018/978-1-60566-064-6.ch017

Saekow, A., & Boonmee, C. (2011). The challenges of implementing e-government interoperability in Thailand: Case of official electronic correspondence letters exchange across government departments. In Y. Charalabidis (Ed.), *Interoperability in digital public services and administration: Bridging e-government and e-business* (pp. 40–61). Hershey, PA: IGI Global. doi:10.4018/978-1-61520-887-6.ch003

Saekow, A., & Boonmee, C. (2012). The challenges of implementing e-government interoperability in Thailand: Case of official electronic correspondence letters exchange across government departments. In *Digital democracy: Concepts, methodologies, tools, and applications* (pp. 1883–1905). Hershey, PA: IGI Global. doi:10.4018/978-1-4666-1740-7.ch094

Sagsan, M., & Medeni, T. (2012). Understanding "knowledge management (KM) paradigms" from social media perspective: An empirical study on discussion group for KM at professional networking site. In M. Cruz-Cunha, P. Gonçalves, N. Lopes, E. Miranda, & G. Putnik (Eds.), *Handbook of research on business social networking: Organizational, managerial, and technological dimensions* (pp. 738–755). Hershey, PA: IGI Global. doi:10.4018/978-1-61350-168-9.ch039

Sahi, G., & Madan, S. (2013). Information security threats in ERP enabled e-governance: Challenges and solutions. In *Enterprise resource planning: Concepts, methodologies, tools, and applications* (pp. 825–837). Hershey, PA: IGI Global. doi:10.4018/978-1-4666-4153-2.ch048

Sanford, C., & Bhattacherjee, A. (2008). IT implementation in a developing country municipality: A sociocognitive analysis. *International Journal of Technology and Human Interaction, 4*(3), 68–93. doi:10.4018/jthi.2008070104

Schelin, S. H. (2003). E-government: An overview. In G. Garson (Ed.), *Public information technology: Policy and management issues* (pp. 120–138). Hershey, PA: IGI Global. doi:10.4018/978-1-59140-060-8.ch006

Schelin, S. H. (2004). Training for digital government. In A. Pavlichev & G. Garson (Eds.), *Digital government: Principles and best practices* (pp. 263–275). Hershey, PA: IGI Global. doi:10.4018/978-1-59140-122-3.ch017

Schelin, S. H. (2007). E-government: An overview. In G. Garson (Ed.), *Modern public information technology systems: Issues and challenges* (pp. 110–126). Hershey, PA: IGI Global. doi:10.4018/978-1-59904-051-6.ch006

Schelin, S. H., & Garson, G. (2004). Theoretical justification of critical success factors. In G. Garson & S. Schelin (Eds.), *IT solutions series: Humanizing information technology: Advice from experts* (pp. 4–15). Hershey, PA: IGI Global. doi:10.4018/978-1-59140-245-9.ch002

Scime, A. (2002). Information systems and computer science model curricula: A comparative look. In M. Dadashzadeh, A. Saber, & S. Saber (Eds.), *Information technology education in the new millennium* (pp. 146–158). Hershey, PA: IGI Global. doi:10.4018/978-1-931777-05-6.ch018

Scime, A. (2009). Computing curriculum analysis and development. In M. Khosrow-Pour (Ed.), *Encyclopedia of information science and technology* (2nd ed.; pp. 667–671). Hershey, PA: IGI Global. doi:10.4018/978-1-60566-026-4.ch108

Scime, A., & Wania, C. (2008). Computing curricula: A comparison of models. In C. Van Slyke (Ed.), *Information communication technologies: Concepts, methodologies, tools, and applications* (pp. 1270–1283). Hershey, PA: IGI Global. doi:10.4018/978-1-59904-949-6.ch088

Seidman, S. B. (2009). An international perspective on professional software engineering credentials. In H. Ellis, S. Demurjian, & J. Naveda (Eds.), *Software engineering: Effective teaching and learning approaches and practices* (pp. 351–361). Hershey, PA: IGI Global. doi:10.4018/978-1-60566-102-5.ch018

Seifert, J. W. (2007). E-government act of 2002 in the United States. In A. Anttiroiko & M. Malkia (Eds.), *Encyclopedia of digital government* (pp. 476–481). Hershey, PA: IGI Global. doi:10.4018/978-1-59140-789-8.ch072

Seifert, J. W., & Relyea, H. C. (2008). E-government act of 2002 in the United States. In A. Anttiroiko (Ed.), *Electronic government: Concepts, methodologies, tools, and applications* (pp. 154–161). Hershey, PA: IGI Global. doi:10.4018/978-1-59904-947-2.ch013

Seufert, S. (2002). E-learning business models: Framework and best practice examples. In M. Raisinghani (Ed.), *Cases on worldwide e-commerce: Theory in action* (pp. 70–94). Hershey, PA: IGI Global. doi:10.4018/978-1-930708-27-3.ch004

Shareef, M. A., & Archer, N. (2012). E-government service development. In M. Shareef, N. Archer, & S. Dutta (Eds.), *E-government service maturity and development: Cultural, organizational and technological perspectives* (pp. 1–14). Hershey, PA: IGI Global. doi:10.4018/978-1-60960-848-4.ch001

Shareef, M. A., & Archer, N. (2012). E-government initiatives: Review studies on different countries. In M. Shareef, N. Archer, & S. Dutta (Eds.), *E-government service maturity and development: Cultural, organizational and technological perspectives* (pp. 40–76). Hershey, PA: IGI Global. doi:10.4018/978-1-60960-848-4.ch003

Shareef, M. A., Kumar, U., & Kumar, V. (2011). E-government development: Performance evaluation parameters. In M. Shareef, V. Kumar, U. Kumar, & Y. Dwivedi (Eds.), *Stakeholder adoption of e-government services: Driving and resisting factors* (pp. 197–213). Hershey, PA: IGI Global. doi:10.4018/978-1-60960-601-5.ch010

Shareef, M. A., Kumar, U., Kumar, V., & Niktash, M. (2012). Electronic-government vision: Case studies for objectives, strategies, and initiatives. In M. Shareef, N. Archer, & S. Dutta (Eds.), *E-government service maturity and development: Cultural, organizational and technological perspectives* (pp. 15–39). Hershey, PA: IGI Global. doi:10.4018/978-1-60960-848-4.ch002

Shukla, P., Kumar, A., & Anu Kumar, P. B. (2013). Impact of national culture on business continuity management system implementation. *International Journal of Risk and Contingency Management, 2*(3), 23–36. doi:10.4018/ijrcm.2013070102

Shulman, S. W. (2007). The federal docket management system and the prospect for digital democracy in U S rulemaking. In G. Garson (Ed.), *Modern public information technology systems: Issues and challenges* (pp. 166–184). Hershey, PA: IGI Global. doi:10.4018/978-1-59904-051-6.ch009

Simonovic, S. (2007). Problems of offline government in e-Serbia. In A. Anttiroiko & M. Malkia (Eds.), *Encyclopedia of digital government* (pp. 1342–1351). Hershey, PA: IGI Global. doi:10.4018/978-1-59140-789-8.ch205

Simonovic, S. (2008). Problems of offline government in e-Serbia. In A. Anttiroiko (Ed.), *Electronic government: Concepts, methodologies, tools, and applications* (pp. 2929–2942). Hershey, PA: IGI Global. doi:10.4018/978-1-59904-947-2.ch219

Singh, A. M. (2005). Information systems and technology in South Africa. In M. Khosrow-Pour (Ed.), *Encyclopedia of information science and technology* (pp. 1497–1502). Hershey, PA: IGI Global. doi:10.4018/978-1-59140-553-5.ch263

Singh, S., & Naidoo, G. (2005). Towards an e-government solution: A South African perspective. In W. Huang, K. Siau, & K. Wei (Eds.), *Electronic government strategies and implementation* (pp. 325–353). Hershey, PA: IGI Global. doi:10.4018/978-1-59140-348-7.ch014

Snoke, R., & Underwood, A. (2002). Generic attributes of IS graduates: An analysis of Australian views. In F. Tan (Ed.), *Advanced topics in global information management* (Vol. 1, pp. 370–384). Hershey, PA: IGI Global. doi:10.4018/978-1-930708-43-3.ch023

Sommer, L. (2006). Revealing unseen organizations in higher education: A study framework and application example. In A. Metcalfe (Ed.), *Knowledge management and higher education: A critical analysis* (pp. 115–146). Hershey, PA: IGI Global. doi:10.4018/978-1-59140-509-2.ch007

Song, H., Kidd, T., & Owens, E. (2011). Examining technological disparities and instructional practices in English language arts classroom: Implications for school leadership and teacher training. In L. Tomei (Ed.), *Online courses and ICT in education: Emerging practices and applications* (pp. 258–274). Hershey, PA: IGI Global. doi:10.4018/978-1-60960-150-8.ch020

Speaker, P. J., & Kleist, V. F. (2003). Using information technology to meet electronic commerce and MIS education demands. In A. Aggarwal (Ed.), *Web-based education: Learning from experience* (pp. 280–291). Hershey, PA: IGI Global. doi:10.4018/978-1-59140-102-5.ch017

Spitler, V. K. (2007). Learning to use IT in the workplace: Mechanisms and masters. In M. Mahmood (Ed.), *Contemporary issues in end user computing* (pp. 292–323). Hershey, PA: IGI Global. doi:10.4018/978-1-59140-926-7.ch013

Stellefson, M. (2011). Considerations for marketing distance education courses in health education: Five important questions to examine before development. In U. Demiray & S. Sever (Eds.), *Marketing online education programs: Frameworks for promotion and communication* (pp. 222–234). Hershey, PA: IGI Global. doi:10.4018/978-1-60960-074-7.ch014

Straub, D. W., & Loch, K. D. (2006). Creating and developing a program of global research. *Journal of Global Information Management, 14*(2), 1–28. doi:10.4018/jgim.2006040101

Straub, D. W., Loch, K. D., & Hill, C. E. (2002). Transfer of information technology to the Arab world: A test of cultural influence modeling. In M. Dadashzadeh (Ed.), *Information technology management in developing countries* (pp. 92–134). Hershey, PA: IGI Global. doi:10.4018/978-1-931777-03-2.ch005

Straub, D. W., Loch, K. D., & Hill, C. E. (2003). Transfer of information technology to the Arab world: A test of cultural influence modeling. In F. Tan (Ed.), *Advanced topics in global information management* (Vol. 2, pp. 141–172). Hershey, PA: IGI Global. doi:10.4018/978-1-59140-064-6.ch009

Suki, N. M., Ramayah, T., Ming, M. K., & Suki, N. M. (2013). Factors enhancing employed job seekers intentions to use social networking sites as a job search tool. In A. Mesquita (Ed.), *User perception and influencing factors of technology in everyday life* (pp. 265–281). Hershey, PA: IGI Global. doi:10.4018/978-1-4666-1954-8.ch018

Suomi, R. (2006). Introducing electronic patient records to hospitals: Innovation adoption paths. In T. Spil & R. Schuring (Eds.), *E-health systems diffusion and use: The innovation, the user and the use IT model* (pp. 128–146). Hershey, PA: IGI Global. doi:10.4018/978-1-59140-423-1.ch008

Swim, J., & Barker, L. (2012). Pathways into a gendered occupation: Brazilian women in IT. *International Journal of Social and Organizational Dynamics in IT, 2*(4), 34–51. doi:10.4018/ijsodit.2012100103

Tarafdar, M., & Vaidya, S. D. (2006). Adoption and implementation of IT in developing nations: Experiences from two public sector enterprises in India. In M. Khosrow-Pour (Ed.), *Cases on information technology planning, design and implementation* (pp. 208–233). Hershey, PA: IGI Global. doi:10.4018/978-1-59904-408-8.ch013

Tarafdar, M., & Vaidya, S. D. (2008). Adoption and implementation of IT in developing nations: Experiences from two public sector enterprises in India. In G. Garson & M. Khosrow-Pour (Eds.), *Handbook of research on public information technology* (pp. 905–924). Hershey, PA: IGI Global. doi:10.4018/978-1-59904-857-4.ch076

Thesing, Z. (2007). Zarina thesing, pumpkin patch. In M. Hunter (Ed.), *Contemporary chief information officers: Management experiences* (pp. 83–94). Hershey, PA: IGI Global. doi:10.4018/978-1-59904-078-3.ch007

Thomas, J. C. (2004). Public involvement in public administration in the information age: Speculations on the effects of technology. In M. Malkia, A. Anttiroiko, & R. Savolainen (Eds.), *eTransformation in governance: New directions in government and politics* (pp. 67–84). Hershey, PA: IGI Global. doi:10.4018/978-1-59140-130-8.ch004

Treiblmaier, H., & Chong, S. (2013). Trust and perceived risk of personal information as antecedents of online information disclosure: Results from three countries. In F. Tan (Ed.), *Global diffusion and adoption of technologies for knowledge and information sharing* (pp. 341–361). Hershey, PA: IGI Global. doi:10.4018/978-1-4666-2142-8.ch015

van Grembergen, W., & de Haes, S. (2008). IT governance in practice: Six case studies. In W. van Grembergen & S. De Haes (Eds.), *Implementing information technology governance: Models, practices and cases* (pp. 125–237). Hershey, PA: IGI Global. doi:10.4018/978-1-59904-924-3.ch004

van Os, G., Homburg, V., & Bekkers, V. (2013). Contingencies and convergence in European social security: ICT coordination in the back office of the welfare state. In M. Cruz-Cunha, I. Miranda, & P. Gonçalves (Eds.), *Handbook of research on ICTs and management systems for improving efficiency in healthcare and social care* (pp. 268–287). Hershey, PA: IGI Global. doi:10.4018/978-1-4666-3990-4.ch013

Velloso, A. B., Gassenferth, W., & Machado, M. A. (2012). Evaluating IBMEC-RJ's intranet usability using fuzzy logic. In M. Cruz-Cunha, P. Gonçalves, N. Lopes, E. Miranda, & G. Putnik (Eds.), *Handbook of research on business social networking: Organizational, managerial, and technological dimensions* (pp. 185–205). Hershey, PA: IGI Global. doi:10.4018/978-1-61350-168-9.ch010

Villablanca, A. C., Baxi, H., & Anderson, K. (2009). Novel data interface for evaluating cardiovascular outcomes in women. In A. Dwivedi (Ed.), *Handbook of research on information technology management and clinical data administration in healthcare* (pp. 34–53). Hershey, PA: IGI Global. doi:10.4018/978-1-60566-356-2.ch003

Villablanca, A. C., Baxi, H., & Anderson, K. (2011). Novel data interface for evaluating cardiovascular outcomes in women. In *Clinical technologies: Concepts, methodologies, tools and applications* (pp. 2094–2113). Hershey, PA: IGI Global. doi:10.4018/978-1-60960-561-2.ch806

Virkar, S. (2011). Information and communication technologies in administrative reform for development: Exploring the case of property tax systems in Karnataka, India. In J. Steyn, J. Van Belle, & E. Mansilla (Eds.), *ICTs for global development and sustainability: Practice and applications* (pp. 127–149). Hershey, PA: IGI Global. doi:10.4018/978-1-61520-997-2.ch006

Virkar, S. (2013). Designing and implementing e-government projects: Actors, influences, and fields of play. In S. Saeed & C. Reddick (Eds.), *Human-centered system design for electronic governance* (pp. 88–110). Hershey, PA: IGI Global. doi:10.4018/978-1-4666-3640-8.ch007

Wallace, A. (2009). E-justice: An Australian perspective. In A. Martínez & P. Abat (Eds.), *E-justice: Using information communication technologies in the court system* (pp. 204–228). Hershey, PA: IGI Global. doi:10.4018/978-1-59904-998-4.ch014

Wang, G. (2012). E-democratic administration and bureaucratic responsiveness: A primary study of bureaucrats' perceptions of the civil service e-mail box in Taiwan. In K. Kloby & M. D'Agostino (Eds.), *Citizen 2.0: Public and governmental interaction through web 2.0 technologies* (pp. 146–173). Hershey, PA: IGI Global. doi:10.4018/978-1-4666-0318-9.ch009

Wangpipatwong, S., Chutimaskul, W., & Papasratorn, B. (2011). Quality enhancing the continued use of e-government web sites: Evidence from e-citizens of Thailand. In V. Weerakkody (Ed.), *Applied technology integration in governmental organizations: New e-government research* (pp. 20–36). Hershey, PA: IGI Global. doi:10.4018/978-1-60960-162-1.ch002

Wedemeijer, L. (2006). Long-term evolution of a conceptual schema at a life insurance company. In M. Khosrow-Pour (Ed.), *Cases on database technologies and applications* (pp. 202–226). Hershey, PA: IGI Global. doi:10.4018/978-1-59904-399-9.ch012

Whybrow, E. (2008). Digital access, ICT fluency, and the economically disadvantages: Approaches to minimize the digital divide. In F. Tan (Ed.), *Global information technologies: Concepts, methodologies, tools, and applications* (pp. 1409–1422). Hershey, PA: IGI Global. doi:10.4018/978-1-59904-939-7.ch102

Whybrow, E. (2008). Digital access, ICT fluency, and the economically disadvantages: Approaches to minimize the digital divide. In C. Van Slyke (Ed.), *Information communication technologies: Concepts, methodologies, tools, and applications* (pp. 764–777). Hershey, PA: IGI Global. doi:10.4018/978-1-59904-949-6.ch049

Wickramasinghe, N., & Geisler, E. (2010). Key considerations for the adoption and implementation of knowledge management in healthcare operations. In M. Saito, N. Wickramasinghe, M. Fuji, & E. Geisler (Eds.), *Redesigning innovative healthcare operation and the role of knowledge management* (pp. 125–142). Hershey, PA: IGI Global. doi:10.4018/978-1-60566-284-8.ch009

Wickramasinghe, N., & Geisler, E. (2012). Key considerations for the adoption and implementation of knowledge management in healthcare operations. In *Organizational learning and knowledge: Concepts, methodologies, tools and applications* (pp. 1316–1328). Hershey, PA: IGI Global. doi:10.4018/978-1-60960-783-8.ch405

Wickramasinghe, N., & Goldberg, S. (2007). A framework for delivering m-health excellence. In L. Al-Hakim (Ed.), *Web mobile-based applications for healthcare management* (pp. 36–61). Hershey, PA: IGI Global. doi:10.4018/978-1-59140-658-7.ch002

Wickramasinghe, N., & Goldberg, S. (2008). Critical success factors for delivering m-health excellence. In N. Wickramasinghe & E. Geisler (Eds.), *Encyclopedia of healthcare information systems* (pp. 339–351). Hershey, PA: IGI Global. doi:10.4018/978-1-59904-889-5.ch045

Wyld, D. (2009). Radio frequency identification (RFID) technology. In J. Symonds, J. Ayoade, & D. Parry (Eds.), *Auto-identification and ubiquitous computing applications* (pp. 279–293). Hershey, PA: IGI Global. doi:10.4018/978-1-60566-298-5.ch017

Yaghmaei, F. (2010). Understanding computerised information systems usage in community health. In J. Rodrigues (Ed.), *Health information systems: Concepts, methodologies, tools, and applications* (pp. 1388–1399). Hershey, PA: IGI Global. doi:10.4018/978-1-60566-988-5.ch088

Yee, G., El-Khatib, K., Korba, L., Patrick, A. S., Song, R., & Xu, Y. (2005). Privacy and trust in e-government. In W. Huang, K. Siau, & K. Wei (Eds.), *Electronic government strategies and implementation* (pp. 145–190). Hershey, PA: IGI Global. doi:10.4018/978-1-59140-348-7.ch007

Yeh, S., & Chu, P. (2010). Evaluation of e-government services: A citizen-centric approach to citizen e-complaint services. In C. Reddick (Ed.), *Citizens and e-government: Evaluating policy and management* (pp. 400–417). Hershey, PA: IGI Global. doi:10.4018/978-1-61520-931-6.ch022

Young-Jin, S., & Seang-tae, K. (2008). E-government concepts, measures, and best practices. In A. Anttiroiko (Ed.), *Electronic government: Concepts, methodologies, tools, and applications* (pp. 32–57). Hershey, PA: IGI Global. doi:10.4018/978-1-59904-947-2.ch004

Yun, H. J., & Opheim, C. (2012). New technology communication in American state governments: The impact on citizen participation. In K. Bwalya & S. Zulu (Eds.), *Handbook of research on e-government in emerging economies: Adoption, e-participation, and legal frameworks* (pp. 573–590). Hershey, PA: IGI Global. doi:10.4018/978-1-4666-0324-0.ch029

Zhang, N., Guo, X., Chen, G., & Chau, P. Y. (2011). User evaluation of e-government systems: A Chinese cultural perspective. In F. Tan (Ed.), *International enterprises and global information technologies: Advancing management practices* (pp. 63–84). Hershey, PA: IGI Global. doi:10.4018/978-1-60960-605-3.ch004

Zuo, Y., & Hu, W. (2011). Trust-based information risk management in a supply chain network. In J. Wang (Ed.), *Supply chain optimization, management and integration: Emerging applications* (pp. 181–196). Hershey, PA: IGI Global. doi:10.4018/978-1-60960-135-5.ch013

# Compilation of References


Abada, I., Briat, V., & Massol, O. (2013). Construction of a fuel demand function portraying interfuel substitution, a system dynamics approach. *Energy*, *49*, 240–251. doi:10.1016/j.energy.2012.10.063

Abo-Hamad, W., & Arisha, A. (2013). Simulation-based framework to improve patient experience in an emergency department. *European Journal of Operational Research*, *224*(1), 154–166. doi:10.1016/j.ejor.2012.07.028

Aggarwal, V. A., Siggelkow, N., & Singh, H. (2011). Governing collaborative activity: Interdependence and the impact of coordination and exploration. *Strategic Management Journal*, *32*(7), 705–730. doi:10.1002/smj.900

Ahmedvand & Arab. (2009). *Application of system dynamics methodology in police*. Tehran: Organizational Development of Police.

Ahrweiler, P., Pyka, A., & Gilbert, N. (2011). A new model for university-industry links in knowledge-based economies. *Journal of Product Innovation Management*, *28*(2), 218–235. doi:10.1111/j.1540-5885.2010.00793.x

Akay, V. (2006). *Business is the people and people are the business: Humanizing the corporation*. iUniverse.

AlDurgham, M. M., & Barghash, M. A. (2008). A generalised framework for simulation-based decision support for manufacturing. *Production Planning and Control*, *19*(5), 518–534. doi:10.1080/09537280802187626

Altiok, T., & Melamed, B. (2000). *Simulation modeling and analysis with ARENA*. Elsevier.

Anderson, P. H., Beveridge, D. A., & Scott, T. W. (2000). *Threshold Entrepreneur. A New Business Simulation*. Upper Saddle River, NJ: Prentice-Hall Inc.

Anton, J. (1999). *Call center performance enhancement. Using simulation and modeling*. Purdue University Press.

AORN. (2012). Recommended Practices for Sterilization in the Perioperative Practice Setting. In Perioperative Standards and Recommended Practices. Denver, CO: AORN.

Arasti, M. R., & Packniat, M. (2011). *A Classification of Models for Technology Strategy Formulation*. Retrieved October 24, 2014, from http://www.nrisp.ac.ir/jstp/browse.php?a_code=A-10-28-64&sid=1&slc_lang=en

Araujo, C. A. S. (2005). *Fatores a serem gerenciados para o alcance da qualidade para os clientes internos: um estudo em um conjunto de hospitais brasileiros*. Rio de Janeiro: Tese de Doutorado em Administração. Universidade Federal do Rio de Janeiro.

Armstrong, M., & Baron, A. (2003). *Strategic HRM the key to improve business performance*. CIPD.

Baker, M. J. (2001). *Marketing. Critical perspectives on business and management*. Routledge.

Banks, J. (1996). *Discrete-Event System Simulation*. Englewood Cliffs, NJ: Prentice-Hall.

Barjis, J., & Pergl, R. (2014). Enterprise and organizational modeling and simulation. *10th In. Workshop, EOMAS 2014*.

Barlas, Y., & Yasarcan, H. (2006). Goal setting, evaluation, learning and revision: A dynamic modeling approach. *Evaluation and Program Planning*, 29.

Barney, J. (1991). Firm resources and sustained competitive advantage. *Journal of Management*, *17*(1), 99–120. doi:10.1177/014920639101700108

Baron, A., & Armstrong, M. (2007). *Human capital management*. Kogan Page Pub.

Barr, P. S., Stimpert, J. L., & Huff, A. S. (1992). Cognitive change, strategic action, and organizational renewal. *Strategic Management Journal*, *13*(S1), 15–36. doi:10.1002/smj.4250131004

Baruch, Y., & Yuen, Y. K. J. (2000). Inclination to opt for teleworking. *International Journal of Manpower*, *21*(7), 521–539. doi:10.1108/01437720010378980

Bashner, D. B. (2014). *Employee benefits design and compensation*. FT Press.

Basu, D., Bhattacharya, S., Mahajan, A., Ramanan, V. R., & Chandy, M. (2015). Sterilization indicators in central sterile supply department: quality assurance and cost implications. *Infection Control & Hospital Epidemiology, 36*(4), 484-486.

Berry, L. L., Parasuraman, A., & Zeithaml, V. A. (1994). Improving Service Quality in America. *The Academy of Management Executive Journal of Marketing*, *8*(2), 31–46.

Bierly, P. E. III, & Coombs, J. E. (2004). Equity alliances, stages of product development, and alliance instability. *Journal of Engineering and Technology Management*, *21*(3), 191–214. doi:10.1016/j.jengtecman.2004.05.001

Bohlander, G., & Snell, S. (2010). *Managing human resources*. Cengage Learning.

Bohn, R. E. (1994). Measuring and managing technological knowledge. *Sloan Management Review*, *36*, 61–73.

Boxall, P. F., & Purcell, J. (2007). *The Oxford handbook of HRM*. Oxford University Press.

Brailsford, S. C., Harper, P. R., Patel, B., & Pitt, M. (2009). An analysis of the academic literature on simulation and modelling in health care. *Journal of Simulation*, *3*(3), 130–140. doi:10.1057/jos.2009.10

Brasil. (2012). *Ministério da Saúde. Agência Nacional de Vigilância Sanitária (ANVISA), Resolução da Diretoria Colegiada (RDC) n. 15 de 15 março de 2012*. Diário Oficial da República Federativa do Brasil, Brasília.

Brasil. Constituição. (1988). *Constituição da República Federativa do Brasil*. Brasília, DF: Senado Federal: Centro Gráfico.

BRAZIL. (2002). *Ministério da Saúde.Agência Nacional de Vigilância Sanitária (ANVISA), Resolução da Diretoria Colegiada (RDC), n. 50 de 21 de fevereiro de 2002*. Brasília: Diário Oficial da República Federativa do Brasil.

Brewster, C., & Larsen, H. H. (2007). *Human resources management in Northern Europe. Trends, dilemas and strategies*. Blackwell Business.

Bridgeland, D. M., & Zahavi, R. (2008). *Business modeling: a practical guide to realizing business value*. Morgan Kaufmann.

Brocklehurst, M. (2001). Power, identity and new technology in homework: Implications for new forms of organizing. *Organization Studies*, *22*(3), 445–466. doi:10.1177/0170840601223003

Bui, T., & Lee, A. N. D. (1999). An agent-based framework for building decision support systems. *Decision Support Systems*, *25*(3), 225–237. doi:10.1016/S0167-9236(99)00008-1

Burch, S. (1991). *Teleworking: a Strategic Guide for Management*. London: Kogan Page.

Burton, R. M., & Obel, B. (2011). Computational modeling for what-is, what-might-be, and what-should-be studies-and triangulation. *Organization Science*, *22*(5), 1195–1202. doi:10.1287/orsc.1100.0635

Buytendijk, F., Hatch, T., & Micheli, P. (2010). Scenario-based strategy maps. *Business Horizons*, *53*(4), 335–347. doi:10.1016/j.bushor.2010.02.002

Caillou, P., Gaudou, B., Grignard, A., Truong, C. Q., & Taillandier, P. (2015). *A simple-to-use BDI architecture for agent-based modeling and simulation*. The Eleventh Conference of the European Social Simulation Association (ESSA 2015), Groningen, Netherlands. Retrieved August 29, 2017, from https://hal-univ-tlse3.archives-ouvertes.fr/hal-01216165/document

Cantens, T., Raballand, G., & Bilangna, S. (2010). *Reforming Customs by measuring performance: a Cameroon case study*. World Customs Journal.

Capkun, V., Messner, M., & Rissbacher, C. (2012). Service specialization and operational performance in hospitals. *International Journal of Operations & Production Management*, *32*(4), 468–495. doi:10.1108/01443571211223103

Carayannis, E. G., Provance, M., & Givens, N. (2011). Knowledge arbitrage, serendipity, and acquisition formality: Their effects on sustainable entrepreneurial activity in regions. *IEEE Transactions on Engineering Management*, *58*(3), 564–577. doi:10.1109/TEM.2011.2109725

Cardoen, B., Demeulemeester, E., & Beliën, J. (2010). Operating room planning and scheduling: A classification scheme. *International Journal of Social Health Information Management*, *1*(1), 71–83.

Cardullo, M. W. (1996). *Introduction to managing technology*. New York: John Wiley & Sons.

Carvalho, H., Barroso, A. P., Machado, V. H., Azevedo, S., & Cruz-Machado, V. (2012). Supply chain redesign for resilience using simulation. *Computers & Industrial Engineering*, *62*(1), 329–341. doi:10.1016/j.cie.2011.10.003

Cassettari, L., Mosca, M., Mosca, R., & Rolando, F. (2013). An healthcare process reengineering using discrete event simulation. *Proceedings of World Congress on Engineering and Computer Science*, 2, 1174-1179.

Chang, M. (2011). *Monte Carlo Simulation for the Pharmaceutical industry. Concepts, algorithms and care studies*. CRC Press.

Cheng, Y.-T., & van de Ven, A. (1996). Learning the innovation journey: Order out of chaos? *Organization Science*, *7*(6), 593–614. doi:10.1287/orsc.7.6.593

Chermack, T. J. (2005). Studying scenario planning: Theory, research suggestions, and hypotheses. *Technological Forecasting and Social Change*, *72*(1), 59–73. doi:10.1016/S0040-1625(03)00137-9

Chesbrough, H. (2006). Open innovation: a new paradigm for understanding industrial innovation. In H. Chesbrough, W. Vanhaverbeke, & J. West (Eds.), Open Innovation: Researching a New Paradigm. Oxford University Press.

Chesbrough, H. (2004). Managing open innovation. *Research Technology Management*, *47*(1), 23–26.

Chetouane, F., Barker, K., & Oropeza, A. S. V. (2012). Sensitivity analysis for simulation-based decision making: Application to a hospital emergency service design. *Simulation Modelling Practice and Theory*, *20*(1), 99–111. doi:10.1016/j.simpat.2011.09.003

Chiesa, V. (1996). Towards a new framework for technology strategy formulation. *Proceedings of the 3rd International Product Development Conference,* 165-178.

Chiesa, V. (2001). *R&D Strategy and Organization*. London: Imperial College Press. doi:10.1142/p223

Chwif, L. (1999). Utilizando simulação de eventos discretos em projetos de sistemas automatizados de manufatura. São Paulo: Academic Press.

Coen, C. A., & Maritan, C. A. (2011). Investing in capabilities: The dynamics of resource allocation. *Organization Science*, *22*(1), 99–117. doi:10.1287/orsc.1090.0524

Coleman Selden, S. (2008). *Human capital*. CQ Press.

Cowan, R., Jonard, N., & Zimmermann, J. B. (2006). Evolving networks of inventors. *Journal of Evolutionary Economics*, *16*(1-2), 155–174. doi:10.1007/s00191-005-0013-1

Cowan, R., Jonard, N., & Zimmermann, J. B. (2007). Bilateral collaboration and the emergence of innovation networks. *Management Science*, *53*(7), 1051–1067. doi:10.1287/mnsc.1060.0618

Cui, A. S. (2013). Portfolio dynamics and alliance termination: The contingent role of resource dissimilarity. *Journal of Marketing*, *77*(3), 15–32. doi:10.1509/jm.11.0164

Cui, A. S., Calantone, R. J., & Griffith, D. A. (2011). Strategic change and termination of interfirm partnerships. *Strategic Management Journal*, *32*(4), 402–423. doi:10.1002/smj.881

Das, T. K., & Teng, B.-S. (2000). Instabilities of strategic alliances: An internal tensions perspective. *Organization Science*, *11*(1), 77–101. doi:10.1287/orsc.11.1.77.12570

Davis, G., & Pecar, B. (2013). *Quantitative methods for decision making using Excel*. Oxford University Press.

DeCenzo, D. A. (1988). *Personnel/Human resouce management*. Englewood Cliffs, NJ: Prentice-Hall.

Demeulemeester, E., Beliën, J., Cardoen, B., & Samudra, M. (2013). Operating room planning and scheduling. In *Handbook of healthcare operations management* (pp. 121–152). Springer New York. doi:10.1007/978-1-4614-5885-2_5

Dhanaraj, C., & Beamish, P. W. (2004). Effect of equity ownership on the survival of international joint ventures. *Strategic Management Journal*, *25*(3), 295–305. doi:10.1002/smj.372

Di Mascolo, M., Gouin, A., & Cong, K. N. (2006). Organization of the production of sterile medical devices. *IFAC Proceedings, 39*(3), 35-40.

Di Mascolo, M., & Gouin, A. (2013). A generic simulation model to assess the performance of sterilization services in health establishments. *Health Care Management Science*, *16*(1), 45–61. doi:10.1007/s10729-012-9210-2 PMID:22886097

Dorfman, M. J., & Medanic, J. (2004). Scheduling trains on a railway network using a discrete event model of railway traffic. *Transportation Research Part B: Methodological*, *38*(1), 81–98. doi:10.1016/S0191-2615(03)00006-7

Doroodian, H. (2008). *Applying Dynamic Systems in Balanced Scorecard to Design Plans of Dynamic Strategies of Small and Medium Industries of Iran* (MS Thesis). Science and Technology University, Tehran, Iran.

Doyle, J. K., & Ford, D. N. (1998). Mental models concepts for system dynamics research. *System Dynamics Review*, *14*(1), 1, 3–29. doi:10.1002/(SICI)1099-1727(199821)14:1<3::AID-SDR140>3.0.CO;2-K

Doz, Y. L. (1996). The evolution of cooperation in strategic alliances: Initial conditions or learning processes? *Strategic Management Journal*, *17*(S1), 55–83. doi:10.1002/smj.4250171006

Doz, Y. L., Olk, P. M., & Ring, P. S. (2000). Formation Processes of R&D Consortia: Which Path to Take? Where Does It Lead? *Strategic Management Journal*, *21*(3), 239–266. doi:10.1002/(SICI)1097-0266(200003)21:3<239::AID-SMJ97>3.0.CO;2-K

Duica, A. (2016). Using simulation and six sigma to analyze and improve motivation process of teachers from undergraduate institutions. *VJES*, *0*(0), 94–103.

Duysters, G., Hagedoorn, J., Lemmens, C. (2003). The effect of alliance block membership on innovative performance. *Revue d'Économie Industrielle, 103*, 59-70.

Ebrahimi, M., Moghbel Baerz, A., Khodadad Hosseini, S. H., & Azar, A. (2013). A New Model of Petrochemical Technology Strategic Planning. *International Journal of Business Administration*, *4*(2), 57–73. doi:10.5430/ijba.v4n2p57

Edvardsson, I. R., & Durst, S. (2013). The Benefits of Knowledge Management in Small and Medium-sized Enterprises. *Procedia: Social and Behavioral Sciences*, *81*(28), 351–354. doi:10.1016/j.sbspro.2013.06.441

Enkel, E., Gassmann, O., & Chesbrough, H. (2009). Open R&D and open innovation: Exploring the phenomenon. *R & D Management*, *39*(4), 311–316. doi:10.1111/j.1467-9310.2009.00570.x

Faems, D., van Looy, B., & Debackere, K. (2005). Inter organizational collaboration and innovation: Toward a portfolio approach. *Journal of Product Innovation Management*, *22*(3), 238–250. doi:10.1111/j.0737-6782.2005.00120.x

Fagiolo, G., & Dosi, G. (2003). Exploitation, exploration and innovation in a model of endogenous growth with locally interacting agents. *Structural Change and Economic Dynamics*, *14*(3), 237–273. doi:10.1016/S0954-349X(03)00022-5

Faulkner, D. (2002). *Strategy: critical perspectives on business and management*. Routledge.

Fazziki, A. E., Nouzri, S., & Sadgal, M. (2010). Towards an MDA Based Multi-agent Approach for Information System Development. *International Journal of Digital Society*, *1*(3), 190–198. doi:10.20533/ijds.2040.2570.2010.0024

Feng, Y. (2012). System dynamics modeling for supply chain information sharing. *Physics Procedia, 25*, 1463–1469. doi:10.1016/j.phpro.2012.03.263

Fireman, S. M. (1999). *A model of telecommuting withdrawal: Employee perceptions predicting the reduction or stopping of telework* (Ph.D. diss.). University of Washington.

Fishman, G. S. (2001). *Discrete-event simulation: modeling, programming, and analysis*. Springer Science&Business Media. doi:10.1007/978-1-4757-3552-9

Fishman, G. S. (2013). *Discrete-Event Simulation: Modeling, Programming, and Analysis*. Springer Science&Business Media.

Fleming, L., & Sorenson, O. (2001). Technology as a complex adaptive system: Evidence from patent data. *Research Policy, 30*(7), 1019–1039. doi:10.1016/S0048-7333(00)00135-9

Florea, N. V. (2015). *Improving employees and organization performance using business simulation and modeling. In Management & Marketing* (pp. 367–381). University of Craiova.

Florea, N. V. (2017). *Simulare si modelare in afaceri. Studii de caz*. Bucuresti: Editura Mustang.

Florea, N. V., & Mihai, D. C. (2015). Improving organization performance through human capital development using a regression function and MATLAB. *JOSA, 3*(32), 229–238.

Florea, N. V., & Mihai, D. C. (2017). Predicting employees evaluation performance using simulation and mathematical modeling. *JOSA, 1*(38), 81–94.

Floyd, C. (1998). *Managing technology for corporate success*. Aldershot, UK: Gower.

Forkmann, S., Wang, D., Henneberg, S. C., Naudé, P., & Sutcliffe, A. (2012). Strategic decision making in business relationships: A dyadic agent-based simulation approach. *Industrial Marketing Management, 41*(5), 816–830. doi:10.1016/j.indmarman.2012.06.010

Forrester, J. W. (1961). Industrial Dynamics. Cambridge, MA:MIT Press.

Forrester, J. W., & Senge, P. M. (1979). *Tests for building confidence in system dynamic models*. Retrieved June 13, 2015, from http://clexchange.org/ftp/documents/Roadmaps/RM10/D-2926-7.pdf

Forrester, J. W. (1971). Counterintuitive behavior of social systems. *Technology Review*, *73*, 52–68.

Forrester, J. W. (1992). Policies, decisions and information sources for modeling. *European Journal of Operational Research*, *59*(1), 42–63. doi:10.1016/0377-2217(92)90006-U

Frenken, K., Izquierdo, L. R., & Zeppini, P. (2012). Branching innovation, recombinant innovation, and endogenous technological transitions. *Environmental Innovation and Societal Transitions*, *4*, 25–35. doi:10.1016/j.eist.2012.06.001

Garud, R., Tuertscher, P., & van de Ven, A. (2013). Perspectives on innovation processes. *The Academy of Management Annals*, *7*(1), 775–819. doi:10.1080/19416520.2013.791066

Geisendorf, S. (2011). Internal selection and market selection in economic genetic algorithms. *Journal of Evolutionary Economics*, *21*(5), 817–841. doi:10.1007/s00191-010-0193-1

Gentner, D. (2002). Psychology of mental models. In N. J. Smelser & P. B. Bates (Eds.), *International Encyclopedia of the Social and Behavioral Sciences* (pp. 9683–9687). Amsterdam: Elsevier Science.

Ghazinoory, S., Divsalar, A., & Soofi, A. S. (2009). A new definition and framework for the development of a national technology strategy: The case of nanotechnology for Iran. *Technological Forecasting and Social Change*, *76*(6), 835–848. doi:10.1016/j.techfore.2008.10.004

Gilsing, V., & Nooteboom, B. (2006). Exploration and exploitation in innovation systems: The case of pharmaceutical biotechnology. *Research Policy*, *35*(1), 1–23. doi:10.1016/j.respol.2005.06.007

Gino, F., & Pisano, G. (2008). Toward a theory of behavioral operations. *Manufacturing & Service Operations Management: M & SOM*, *10*(4), 676–691. doi:10.1287/msom.1070.0205

Glick, M. B., Chermack, T. J., Luckel, H., & Gauck, B. Q. (2012). Effects of scenario planning on participant mental models. *European Journal of Training and Development*, *6*(5), 488–50. doi:10.1108/03090591211232066

Goldvarg, E., & Johnson-Laird, P. N. (2001). Naive causality: A mental model theory of causal meaning and reasoning. *Cognitive Science*, *25*(4), 565–610. doi:10.1207/s15516709cog2504_3

Gray, M., Hodson, N., & Gordon, G. (1994). *Teleworking Explained*. Chichester, UK: John Wiley & sons.

Greiner, R., Puig, J., Huchery, C., Collier, N., & Garnett, S. T. (2014). Scenario modeling to support industry strategic planning and decision making. *Environmental Modelling & Software*, *55*, 120–131. doi:10.1016/j.envsoft.2014.01.011

Grünberg, T. (2004). Performance improvement Towards a method for finding and prioritising potential performance improvement areas in manufacturing operations, Management. *Journal of Productivity and Performance International*, *53*(1), 52–71. doi:10.1108/17410400410509969

Guerriero, F., & Guido, R. (2011). Operational research in the management of the operating theatre: A survey. *Health Care Management Science*, *14*(1), 89–114. doi:10.1007/s10729-010-9143-6 PMID:21103939

Gulati, R., Nohria, N., & Zaheer, A. (2000). Strategic networks. *Strategic Management Journal*, *21*(3), 203–215. doi:10.1002/(SICI)1097-0266(200003)21:3<203::AID-SMJ102>3.0.CO;2-K

Günal, M. M., & Pidd, M. (2010). Discrete event simulation for performance modelling in health care: A review of the literature. *Journal of Simulation*, *4*(1), 42–51. doi:10.1057/jos.2009.25

Hagedoorn, J., & Duysters, G. (2002). External sources of innovative capabilities: The preference for stratregic alliances or mergers and acquisitions. *Journal of Management Studies*, *39*(2), 167–188. doi:10.1111/1467-6486.00287

Hagedoorn, J., & Frankort, H. T. W. (2008). The gloomy side of embeddedness: The effects of overembeddeddness on inter-firm partnership formation. *Network Strategy*, *25*, 503–530. doi:10.1016/S0742-3322(08)25014-X

Hall, R. W. (2013). *Patient flow: reducing delay in healthcare delivery* (Vol. 206). Springer Science & Business Media. doi:10.1007/978-1-4614-9512-3_1

Hamel, G., Doz, Y.L., & Prahalad, C.K. (1989, Jan.). Collaborate with your competitors and win. *Harvard Business Review*.

Handy, S. L., & Mokhtarian, P. L. (1996). The future of telecommuting. *Future*, *28*(3), 227–240. doi:10.1016/0016-3287(96)00003-1

Harrigan, K. R., & Newman, W. H. (1990). Bases of interorganization co-operation: Propensity, power, persistence. *Journal of Management Studies*, *27*(4), 417–434. doi:10.1111/j.1467-6486.1990.tb00255.x

Harris, M. M. (2012). *Handbook of research of international HRM*. Psychology Press.

Hartanto, F., Kreutzer, W., Pawlikowski, K., & Sirisena, H. R. (1996). Quantitative stochastic simulation of telecommunication networks in DESC++. *Computers & Electrical Engineering*, *22*(6), 367–381. doi:10.1016/S0045-7906(96)00015-8

Hart, C. (1998). Doing a literature review: Releasing the social science research imagination. *Sage (Atlanta, Ga.)*.

Henderson, S. G., & Mason, A. J. (2005). Ambulance service planning: simulation and data visualisation. In Operations research and health care (pp. 77-102). Springer US. doi:10.1007/1-4020-8066-2_4

Henderson-Sellers, B., & Giorgini, P. (2005). *Agent-Oriented Methodologies*. Idea Group Pub. doi:10.4018/978-1-59140-581-8

Hill & Hawkins. (1996). Work and family in the virtual office. *Family Relations, 45*(3).

Hoffmann, B., Häfele, S., & Karl, U. (2013). Analysis of performance losses of thermal power plants in Germany - A System Dynamics model approach using data from regional climate modeling. *Energy*, *49*, 193–203. doi:10.1016/j.energy.2012.10.034

Hoffmann, W. H. (2007). Strategies for managing portfolio of alliances. *Strategic Management Journal*, *28*(8), 827–856. doi:10.1002/smj.607

Holbeche, L., & Mayo, A. (2012). *Motivating people in Lean organizations*. Butterworth-Heineman.

Hollnagel, E., Braithwaite, J., & Wears, R. L. (2013). *Epilogue: How to make health care resilient. In Resilient health care* (pp. 227–238). Surrey, UK: Ashgate Publishing Limited.

Holm, L. B., Lurås, H., & Dahl, F. A. (2013). Improving hospital bed utilisation through simulation and optimisation: With application to a 40% increase in patient volume in a Norwegian General Hospital. *International Journal of Medical Informatics*, *82*(2), 80–89. doi:10.1016/j.ijmedinf.2012.05.006 PMID:22698645

Hopp, W. J., & Lovejoy, W. S. (2012). *Hospital operations: Principles of high efficiency health care*. FT Press.

Houben, G., Lenie, K., & Vanhoof, K. (1999). A Knowledge-based SWOT-analysis System as an Instrument for Strategic Planning in Small and Medium Sized Enterprises. *Decision Support Systems*, *26*(2), 125–135. doi:10.1016/S0167-9236(99)00024-X

Hulshof, P. J., Kortbeek, N., Boucherie, R. J., Hans, E. W., & Bakker, P. J. (2012). Taxonomic classification of planning decisions in health care: A structured review of the state of the art in OR/MS. *Health Systems (Basingstoke, England)*, *1*(2), 129–175. doi:10.1057/hs.2012.18

Huws, U., Korte, W. B., & Robinson, S. (1990). Telework: Towards the Elusive Office. Chichester, UK: John Wiley & Sons.

Illeges, V., & Verbeke, A. (2004). Telework: What dose it mean for management? *Long Range Planning*, 32.

Inkpen, A. C., & Ross, J. (2001). Why do some strategic alliances persist beyond their useful life? *California Management Review*, *44*(1), 132–148. doi:10.2307/41166114

Jacques, J. É. (2007). *Gestão estratégica da criação do conhecimento nas organizações hospitalares: um estudo baseado na construção de protocolos médico-assistenciais*. Academic Press.

Jalali, A. A. (2010). Telework, Should and should not. *The Journal of Information Era Analysts*.

Janssen, M. (2015). *Policy Practice and Digital Science: Integrating Complex Systems, Social Simulation and Public Administration in Policy Research*. Springer. doi:10.1007/978-3-319-12784-2

Katsaliaki, K., & Mustafee, N. (2011). Applications of simulation within the healthcare context. *The Journal of the Operational Research Society*, *62*(8), 1431–1451. doi:10.1057/jors.2010.20

Kauffman, S. (1989). Adaptation on rugged fitness landscapes. In D. L. Stein (Ed.), *Lectures in the Sciences of Complexity*. Reading, MA: Addison-Wesley.

Kelly, M. J., Schaan, J.-L., & Joncas, H. (2002). Managing alliance relationships: Key challenges in the early stages of collaboration. *R & D Management*, *32*(1), 11–22. doi:10.1111/1467-9310.00235

Kelly, M. M. (1985). The next workplace revolution: Telecommuting. *Supervisory Management*, *30*(10).

Kelton, W. D., Sturrock, R. P., Kelton, D. T. D., Sadowski, R. P., & Sturrock, D. T. (2008). *Simulación con software Arena*. McGraw-Hill.

Khani, M. (2011). *Developed Administrative System and Teleworking*. Modir Site.

Khodabakhshian, A., Khosravi, M., & Mashayekhi, A. N. (2013). R&D project portfolio in research and development, part of a whole in R&D strategy. *System Dynamics Conference*. Retrieved August 30, 2017, from http://www.systemdynamics.org/conferences/2013/proceed/papers/P1299.pdf

Kim, S. W., & Choi, K. (2009). *A Dynamic Analysis of Technological Innovation Using System Dynamics*. POMS 20th Annual Conference, Orlando, FL. Retrieved August 30, 2017, from https://www.pomsmeetings.org/ConfPapers/011/011-0622.pdf

Kolkman, M.J., Kok, M., van der Veen, A. (2005). Mental model mapping as a new tool to analyse the use of information in decision-making in integrated water management. *Physics and Chemistry of the Earth*, *30*(4-5), 317–332.

Konrad, R., DeSotto, K., Grocela, A., McAuley, P., Wang, J., Lyons, J., & Bruin, M. (2013). Modeling the impact of changing patient flow processes in an emergency department: Insights from a computer simulation study. *Operations Research for Health Care*, *2*(4), 66–74. doi:10.1016/j.orhc.2013.04.001

Korber, M., Paier, M., & Fischer, M. M. (2009). An agent-based view of the biotech innovation system. *Third Central European Conference in Regional Science*.

Kortea, F. R., & Chermack, T. J. (2007). Changing organizational culture with scenario planning. *Futures*, *39*(6), 645–656. doi:10.1016/j.futures.2006.11.001

Kotler, P., & Armstrong, G. (2009). *Principles of Marketing*. Prentice Hall.

Kowalski, K. B., & Swanson, J. A. (2005). Critical success factors in developing teleworking programs, Benchmarking. *International Journal (Toronto, Ont.)*, *12*(3).

Kubíčková, L., Votoupalová, M., & Toulová, M. (2014). Key Motives for Internationalization Process of Small and Medium–Sized Enterprises. *Procedia Economics and Finance*, *12*, 319–328. doi:10.1016/S2212-5671(14)00351-7

Kusumaningdyah, W., Eunike, A., & Yuniarti, R. (2013). Modeling tradeoff in ship breaking industry considering sustainability aspects: A system dynamics approach. *Procedia Environmental Sciences*, *17*, 785–794. doi:10.1016/j.proenv.2013.02.096

Laguna, M., & Marklund, J. (2011). *Business process modeling, simulation and design*. Pearson Edu.

Lakhdar, S. (2001). *Gestion de ressources humaines*. Universite de Boeck.

Lall, S. (1993). Promoting technology development: The role of technology transfer and indigenous effort. *Third World Quarterly*, *14*(1), 95–109. doi:10.1080/01436599308420315

Lamiri, M., Xie, X., Dolgui, A., & Grimaud, F. (2008). A stochastic model for operating room planning with elective and emergency demand for surgery. *European Journal of Operational Research*, *185*(3), 1026–1037. doi:10.1016/j.ejor.2006.02.057

Lane, D. C. (2000). Should system dynamics be described as a hard or deterministic systems approach. *Systems Research and Behavioral Science System*, *17*(1), 3–22. doi:10.1002/(SICI)1099-1743(200001/02)17:1<3::AID-SRES344>3.0.CO;2-7

Law, A. M., & Kelton, D. W. (1991). *Simulation modeling and analysis*. McGraw-Hill.

Law, A. M., & Kelton, W. D. (2000). *Simulation Modeling and Analysis*. Singapore: McGraw-Hill.

Lawal, B. (2014). *Applied Statistical Methods in Agriculture, Health and Life Sciences*. Springer.

Lazer, D., & Friedman, A. (2007). The network structure of exploration and exploitation. *Administrative Science Quarterly*, *52*(4), 667–694. doi:10.2189/asqu.52.4.667

Lee, C.-F., & Chung, C.-F. (2012). An inventory model for deteriorating items in a supply chain with system dynamics analysis. *Procedia: Social and Behavioral Sciences*, *40*, 41–51. doi:10.1016/j.sbspro.2012.03.159

Leite, F. B. (2008). *Central de Material Esterilizado Projeto de Reestruturação e Ampliação do Hospital Regional de Francisco Sá*. Centro Universitário Euroamericano–UNIEURO.

Lenox, M. J., Rockart, S. F., & Lewin, A. Y. (2006). Interdependency, competition, and the distribution of firm and industry profits. *Management Science*, *52*(5), 757–772. doi:10.1287/mnsc.1050.0495

Leonard-Barton, D. (1995). *Wellsprings of knowledge*. Boston: Harvard Business School Press.

Levinthal, D. A. (1997). Adaptation on rugged landscapes. *Management Science*, *43*(7), 934–950. doi:10.1287/mnsc.43.7.934

Lhuillery, S., & Pfister, E. (2009). R&D cooperation and failures in innovation projects: Empirical evidence from French CIS data. *Research Policy*, *38*(1), 45–57. doi:10.1016/j.respol.2008.09.002

Li, A., & Maani, K. (2011). *Dynamic Decision-Making, Learning and Mental Models*. Retrieved August 10, 2014, from http://www.systemdynamics.org/conferences/2011/proceed/papers/P1189.pdf

Li, S. (2007). AgentStra: An Internet-based multi-agent intelligent system for strategic decision-making. *Expert Systems with Applications*, *33*(3), 565–571. doi:10.1016/j.eswa.2006.05.018

Li, S., & Li, J. Z. (2010). AgentsInternational: Integration of multiple agents, simulation, knowledge bases and fuzzy logic for international marketing decision making. *Expert Systems with Applications*, *37*(3), 2580–2587. doi:10.1016/j.eswa.2009.08.022

Lister, K., & Harnish, T. (2010). *Work shifting benefits: The bottem line*. Telework Research Network/Aerodite Inc.

Litvak, E. (Ed.). (2010). *Managing patient flow in hospitals: Strategies and solutions*. Joint Commission Resources.

Liuguo, S., Shijing, Z., & Jianbai, H. (2012). Pricing simulation platform based on system dynamics. *Systems Engineering Procedia*, *5*, 445–453. doi:10.1016/j.sepro.2012.04.067

Llerena, P., & Ozman, M. (2013). Networks, irreversibility and knowledge creation. *Journal of Evolutionary Economics*, *23*(2), 431–453. doi:10.1007/s00191-011-0231-7

Lunnan, R., & Haugland, S. A. (2008). Predicting and measuring alliance performance: A multidimensional analysis. *Strategic Management Journal*, *29*(5), 545–556. doi:10.1002/smj.660

Maidstone, R. (2012). *Discrete event simulation, system dynamics and agent based simulation: Discussion and comparison*. Retrieved from https://www.researchgate.net/publication/306157680_Discrete_Event_Simulation_System_Dynamics_and_Agent_Based_Simulation_Discussion_and_Comparison

Makino, S., Chan, C. M., Isobe, T., & Beamish, P. W. (2007). Intended and unintended termination of international joint ventures. *Strategic Management Journal*, *28*(11), 1113–1132. doi:10.1002/smj.629

Malagón-Londoño, G., Morera, R. G., Laverde, G. P., & Paulo, A. F. D. (2003). *Administração hospitalar. 2*. Buenos Aires: Editorial Médica Panamericana.

Mallor, F., & Azcárate, C. (2014). Combining optimization with simulation to obtain credible models for intensive care units. *Annals of Operations Research, 221*(1), 255–271. doi:10.1007/s10479-011-1035-8

Manea, M. D. (2012). *Contabilitatea instituţiilor publice*. Bucuresti: Academic Press.

Manolescu, A. (2006). *Managementul resurselor umane*. Bucuresti: Ed. Economica.

Mantoani, C. C. (2012). *A transfusão sanguínea no perioperatório e a ocorrência da infecção de sítio cirúrgico* (Doctoral dissertation). Universidade de São Paulo.

Marcus, G. P. (1992). Present and emergent identities: requirements for ethnographies of late twentieth-century modernity worldwide. In S. Lash & J. Friedman (Eds.), *Modernity and identity* (pp. 309–330). Oxford, UK: Blackwell.

Marques, I., & Captivo, M. E. (2015). Bicriteria elective surgery scheduling using an evolutionary algorithm. *Operations Research for Health Care, 7*, 14–26. doi:10.1016/j.orhc.2015.07.004

Marz, S., Friedrich-Nishio, M., & Grupp, H. (2005). Knowledge transfer in an innovation simulation model. *Technological Forecasting and Social Change, 73*(2), 138–152. doi:10.1016/j.techfore.2005.05.002

Ma, T., & Nakamori, Y. (2005). Agent-based modeling on technological innovation as an evolutionary process. *European Journal of Operational Research, 166*(3), 741–755. doi:10.1016/j.ejor.2004.01.055

Mathis, R., Jackson, J., Valentine, S., & Meglich, P. (2016). *Human resources management*. Cengage Learning.

McCutchen, W. W. Jr, Swamidass, P. M., & Teng, B.-S. (2008). Strategic alliance termination and performance: The role of task complexity, nationality, and experience. *The Journal of High Technology Management Research, 18*(2), 191–202. doi:10.1016/j.hitech.2007.12.001

McHaney, R. (2009). *Understanding computer simulation*. Ventus Pub., ApS.

McLinden, G. (2005). *Integrity in Customs, Customs modernization handbook*. Washington, DC: World Bank.

Milling, P. M., & Maier, F. H. (2009). Research and development, technological innovations and diffusion. *System Dynamics*. Retrieved August 30, 2017, from http://www.eolss.net/sample-chapters/c15/E6-63-01-06.pdf

Millmore, M. (2007). *Strategic HRM: contemporary issues*. Pearson Education.

Mittal, S. (2017). *Guide to Simulation-Based Disciplines: Advancing Our Computational Future*. Springer. doi:10.1007/978-3-319-61264-5

Monga, P. (2001). *A System dynamics model of the development of new technologies for ship systems*. Master of Science in Industrial and Systems Engineering. Retrieved August 30, 2017, from https://theses.lib.vt.edu/theses/available/etd-10022001-014531/unrestricted/Pavinder_Monga_Masters_Thesis.pdf

Moorcreft & Bennett. (1995). *European Guide to Teleworking: A Frame Work for Action.* Dublin, Ireland: European Foundation for the Improvement of Living and Working Conditions.

Moore, J. H., & Weatherford, L. R. (2001). *Decision Modeling with Microsoft Excel*. Prentice-Hall.

Morecroft, J. D. W. (1988). System dynamics and micro worlds for policymakers. *European Journal of Operational Research*, *35*(3), 301–320. doi:10.1016/0377-2217(88)90221-4

Mowery, D. C., Oxley, J. E., & Silverman, B. S. (1996). Strategic alliances and interfirm knowledge transfer. *Strategic Management Journal*, *17*(S2), 77–91. doi:10.1002/smj.4250171108

Moynihan, D. P., & Pandey, S. K. (2007). The role of organizations in fostering public service motivation. *Public Administration Review*, *67*(1), 40–53. doi:10.1111/j.1540-6210.2006.00695.x

Muller, M., Buchmann, T., & Kudic, M. (2014). Micro strategies and macro patterns in the evolution of innovation networks: An agent-based simulation approach. In N. Gilbert, P. Ahrweiler, & A. Pyka (Eds.), *Simulating Knowledge Dynamics in Innovation Networks*. Springer-Verlag Berlin Heidelberg.

Nelson, R. R., & Winter, S. G. (1982). The Schumpeterian tradeoff revisited. *The American Economic Review*, *72*(1), 114–132. doi:10.1007/978-3-662-43508-3_4

Nilles, J. M. (1998). *Managing Telework: Strategies for Managing the Virtual Workforce*. New York: John Wiely & Sons, Inc.

Niu, Q., Peng, Q., El, Y., & Mekkawy, T. (2013). Improvement in the operating room efficiency using Tabu search in simulation. *Business Process Management Journal*, *19*(5), 799–818. doi:10.1108/BPMJ-Nov-2011-0081

Ohori, K., & Takahashi, S. (2012). Market design for standardization problems with agent-based social simulation. *Journal of Evolutionary Economics*, *22*(1), 49–77. doi:10.1007/s00191-010-0196-y

Olk, P., & Young, C. (1997). Why members stay in or leave an R&D consortium: Performance and conditions of membership as determinants of continuity. *Strategic Management Journal*, *18*(11), 855–877. doi:10.1002/(SICI)1097-0266(199712)18:11<855::AID-SMJ923>3.0.CO;2-Q

Padgham, L., Thangarajah, J., & Winikoff, M. (2008). Prometheus design tool. *Proceedings of the Twenty-Third AAAI Conference on Artificial Intelligence*. Retrieved January 13, 2016, from https://www.aaai.org/Papers/AAAI/2008/AAAI08-330.pdf

Palmer, S. (2012). *People an self management*. Butterworth-Heineman.

Pangarkar, N. (2003). Determinants of alliance duration in uncertain environments: The case of the biotechnology sector. *Long Range Planning*, *36*(3), 269–284. doi:10.1016/S0024-6301(03)00041-4

Pangarkar, N. (2009). Do firms learn from alliance terminations? An empirical examination. *Journal of Management Studies*, *46*(6), 982–1004. doi:10.1111/j.1467-6486.2009.00844.x

Pappas, C. (1984). Strategic management of technology. *Journal of Product Innovation Management*, *1*(1), 30–35. doi:10.1016/S0737-6782(84)80040-5

Park, S. H., & Ungson, G. R. (1997). The effect of national culture, organizational complementarity, and economic motivation on joint venture dissolution. *Academy of Management Journal*, *40*(2), 279–307. doi:10.2307/256884

Park, S., Kim, B.-J., & Jung, S.-Y. (2014). Simulation methods of a system dynamics model for efficient operations and planning of capacity expansion of activated-sludge wastewater treatment plants. *Procedia Engineering*, *70*, 1289–1295. doi:10.1016/j.proeng.2014.02.142

Patzelt, H., & Shepherd, D. A. (2008). The decision to persist with underperforming alliances: The role of trust and control. *Journal of Management Studies*, *45*(7), 1218–1243. doi:10.1111/j.1467-6486.2008.00791.x

Paul, J. A., & Lin, L. (2012). Models for improving patient throughput and waiting at hospital emergency departments. *The Journal of Emergency Medicine*, *43*(6), 1119–1126. doi:10.1016/j.jemermed.2012.01.063 PMID:22902245

Pedroso, M. C., & Malik, A. M. (2011). As quatro dimensões competitivas da saúde. *Harvard Business Review*, *89*, 59–67.

Pereira, M. C. O. (2010). *Preparo, embalagem, transporte e armazenagem. In Limpeza, desinfecção e esterilização de artigos em serviços de saúde* (pp. 83–107). São Paulo: APECIH.

Perez, M. (2004). A technology acceptance model of innovation adoption: The case of teleworking. *European Journal of Innovation Management*, *7*(4), 280–291. doi:10.1108/14601060410565038

Perez, M., Sanchez, A. M., Carnicer, M. P. L., & Jimienez, M. J. V. (2004). The environmental impacts of telewoking: A model of urban analysis and a case study. *Management of Environmental Quality*, *15*(6). doi:10.1108/14777830410560728

Perkins, J. J. (1983). *Principles and methods of sterilization in health sciences* (2nd ed.). Charles C. Thomas Pub.

Pezzi, M. C. S., & Leite, J. L. (2010). Investigação em central de material e esterilização utilizando a teoria fundamentada em dados. *Revista Brasileira de Enfermagem*, *63*(3), 391–396. doi:10.1590/S0034-71672010000300007 PMID:20658072

Phillips, J., & Pullian Phillips, P. (2015). *High impact human capital startegy*. AMACOM.

Pierson, K. M. (2006). *Home-based work: A form of telework with advantages and disadvantages related to health outcomes* (Master's thesis). Northern Kentucky University.

Pinson, D. S., Louq, J. A., & Moraitis, P. (1997). A distributed decision support system for strategic planning. *Decision Support Systems*, *20*(1), 35–51. doi:10.1016/S0167-9236(96)00074-7

Pittet, D., Allegranzi, B., Storr, J., Nejad, S. B., Dziekan, G., Leotsakos, A., & Donaldson, L. (2008). Infection control as a major World Health Organization priority for developing countries. *The Journal of Hospital Infection*, *68*(4), 285–292. doi:10.1016/j.jhin.2007.12.013 PMID:18329137

Pophal, L. G. (2002). *Telecommuting: Managing off-site staff for small business*. International Self-Counsel.

Possari, J. F. (2010). *Assistência de enfermagem na recuperação anestésica*. São Paulo: Iátria.

Potiron, K., El Fallah Seghrouchni, A., & Taillibert, P. (2013). *From fault classification to fault tolerance for multi-agent systems*. Springer. Retrieved August 28, 2017, from http://www.springer.com/gp/book/9781447150459

Powell, G. R. (2008). *Marketing calculator. Measuring and managing return on marketing investment*. Singapore: John Wiley & Sons.

Prasad, A. (2011). *Future Internet Services and Service Architectures*. River Pub.

Rasjidin, R., Kumar, A., Alam, F., & Abosuliman, S. (2012). A system dynamics conceptual model on retail electricity supply and demand system to minimize retailer's cost in eastern Australia. *Procedia Engineering*, *49*, 330–337. doi:10.1016/j.proeng.2012.10.145

Reece, B. L. (2011). *Effective human relations. Interpersonal and organizational applications*. South-Western Cengage Learning.

Reed, A. (2001). *Innovation in HRM: tooling up for the talent wars*. London: CIPD House.

Resolução, RDC (2002). Nº 50 de 21 de fevereiro de 2002. Ministério da saúde.

Reuer, J. J., & Zollo, M. (2005). Termination outcomes of research alliances. *Research Policy*, *34*(1), 101–115. doi:10.1016/j.respol.2004.11.003

Reymondon, F., Pellet, B., & Marcon, E. (2008). Optimization of hospital sterilization costs proposing new grouping choices of medical devices into packages. *International Journal of Production Economics*, *112*(1), 326–335. doi:10.1016/j.ijpe.2006.12.066

Robinson, S. (2004). *Simulation. The practice of model development and use*. John Wiley & Sons.

Robson, M. J., Schlegelmilch, B. B., & Bojkowszky, B. (2012). Resource deployment stability and performance in international research-and-development alliances: A self-determination theory explanation. *Journal of International Marketing*, *20*(1), 1–18. doi:10.1509/jim.11.0072

Rodriguez, J. (2012). *Trends in Practical Applications of Agents and Multiagent Systems*. Springer. doi:10.1007/978-3-642-28795-4

Rodriguez-Ulloa, R., & Paucar-Caceres, A. (2004). *Soft system dynamics methodology (SSDM): A combination of soft systems methodology (SSM) and system dynamics (SD)*. Retrieved August 6, 2015, from http://www.systemdynamics.org/conferences/2004/SDS_2004/PAPERS/163PAUCA.pdf

Rooks, G., Snijders, C., & Duysters, G. (2013). Ties that tear apart: The social embeddedness of strategic alliance termination. *The Social Science Journal*, *50*(3), 359–366. doi:10.1016/j.soscij.2013.03.008

Ruth, S., & Chaudhry, I. (2008). *Telework: A productivity paradox? Public Policy*. IEEE Computer Society.

Sa Silva, P. (2011). *Ch.* Simulation, Games and challenges from schools to enterprises. In *Enterprise and organizational modelling and simulation, 7th International Workshop, EOMAS 2011*. Springer.

Sabuncuoglu, I. (2015). *Industrial engineering applications in emerging countries*. CRC Press.

Sadowski, B., & Duysters, G. (2008). Strategic technology alliance termination: An empirical investigation. *Journal of Engineering and Technology Management*, *25*(4), 305–320. doi:10.1016/j.jengtecman.2008.10.002

Sakurada, N., & Miyake, D. I. (2009). Aplicação de simuladores de eventos discretos no processo de modelagem de sistemas de operações de serviços. *Gestão & Produção*, *16*(1), 25–43. doi:10.1590/S0104-530X2009000100004

Saliba, G. (2009). Windows for the mind: The use of scenario planning for enhancing decision-making and managing uncertainty. *Journal of Futures Studies*, *13*(3), 123–129.

Savin, I., & Egbetokun, A. (2016). Emergence of innovation networks from R&D cooperation with endogenous absorptive capacity. *Journal of Economic Dynamics & Control*, *64*, 82–103. doi:10.1016/j.jedc.2015.12.005

Schilling, M. A. (2002). Technology success and failure in winner-take-all markets: The impact of learning orientation, timing, and network externalities. *Academy of Management Journal*, *45*(2), 387–398. doi:10.2307/3069353

Schrempf, B., & Ahrweiler, P. (2012). Modelling innovation networks of general purpose technologies – the case of nanotechnology. *Proceedings of the 2012 Winter Simulation Conference*. doi:10.1109/WSC.2012.6465327

Seavey, R. N. (2007). Unloading, Storage, Distribuition, Transportation and Aseptic Presentation of sterile items. *Managing Infection Control, Denver*, *70*(9), 88–101.

Seddighi, H. (2000). *Econometrics: A Practical Approach*. Psychology Press.

Seddighi, H. (2013). *Introductory Econometrics: A Practical Approach*. Routledge.

Senge, P. M., & Sterman, J. D. (1992). Systems thinking and organizational learning: Acting locally and thinking globally in the organization of the future. *European Journal of Operational Research*, *59*(1), 137–150. doi:10.1016/0377-2217(92)90011-W

Shane, S. (2009). *Technology strategy for managers and entrepreneurs*. Pearson/ Prentice Hall.

Shannon, R. E. (1975). *Systems Simulation: The Art and Science*. Prentice-Hall.

Shim, J. K., & Siegel, J. G. (1999). *Operations management*. Barron's Educational Series.

Siggelkow, N., & Rivkin, J. W. (2006). When exploration backfires: Unintended consequences of multilevel organizational search. *Academy of Management Journal*, *49*(4), 779–795. doi:10.5465/AMJ.2006.22083053

Siha, S. M., & Monroe, R. W. (2006). Telecommuting's past and future: A literature review and research agenda. *Business Process Management Journal*, *12*(4), 455–482. doi:10.1108/14637150610678078

Silva, A. C. P. V., Assad, D. B. N., Oliveira, F. L. C., & Spiegel, T. (2016). Uso de simulação de eventos discretos para redimensionamento de capacidade de centro cirúrgico de um hospital de emergência. In XLVIII Simpósio Brasileiro de Pesquisa Operacional - SBPO 2016. Anais do XLVIII Simpósio Brasileiro de Pesquisa Operacional.

Silva, E. L., & Menezes, E. M. (2001). *Metodologia da pesquisa e elaboração de dissertação*. Florianópolis: Laboratório de Ensino a Distância da UFSC.

Silva, S. B., Caulliraux, H. M., & Spiegel, T. (2015). Modelo de simulação para auxílio em planejamento e gestão de serviços de saúde. *S & G. Sistemas & Gestão*, *10*(4), 608–616. doi:10.20985/1980-5160.2015.v10n4.616

Sims, R. R. (2002). *Organizational success through effective human resources management*. Greenwood Pub.

Slack, N., Chambers, S., & Johnston, R. (2009). *Administração da produção*. Atlas.

Smith, A. C., Barry, R., & Brubaker, C. (2007). *Going lean: Busting barriers to patient flow*. Health Administration Press.

SOBECC. (2013). *Recuperação anestésica e Centro de Material e Esterilização: práticas recomendadas*. São Paulo: Sociedade Brasileira de Enfermeiros de Centro Cirúrgico.

Sohrabi, B. (2003). *Teleworking, Concepts and Planning Process*. Tehran: Irandoc.

Sokolowski, J. A., & Banks, C. M. (2011). *Principles of modeling and simulation. A multidisciplinary approach*. John Wiley & Sons.

SOMASUS. (2011). *Brasil. Ministério da Saúde. Secretaria-Executiva. Departamento de Economia da Saúde e Desenvolvimento. Programação Arquitetônica de Unidades Funcionais de Saúde / Ministério da Saúde, Secretaria-Executiva, Departamento de Economia da Saúde e Desenvolvimento*. Brasília: Ministério da Saúde.

Spiegel, T. (2013). *Contribuições das ciências cognitivas à gestão de operações: análise do impacto da experiência nas decisões do gestor de operações* (Doctoral dissertation). Federal University of Rio de Janeiro.

Spiegel, T. (2014). An Overview of Cognition Roles in Decision-Making. In Encyclopedia of Business Analytics and Optimization. IGI Global. doi:10.4018/978-1-4666-5202-6.ch008

Spiegel, T., & Assad, D. B. N. (2016). Operations Project and Management in Trauma Centers: The Case of Brazilian Units. In Handbook of Research on Healthcare Administration and Management. IGI Global.

Spiegel, T., & Cameira, R. F. (2016). The Processing Chain of Sterilized Material from the Health Units of Rio De Janeiro County: Analysis and Proposition of Scenarios. In Supply Chain Management: Applications for Manufacturing and Service Industries. New York: Nova Science Publishers.

Spiegel, T., Caulliraux, H. M., Treistman, F. E., Guimaraes, G. E., Vaccaro, G. L. R., Amorim, L. G., . . . Azzam, S. M. (2016). Projeto e Gestão de Operações em Saúde. In Tópicos Emergentes e Desafio Metodológicos em Engenharia de Produção: Casos, Experiências e Proposições. Rio de Janeiro: ABEPRO.

Spiegel, T., Vasconcelos, P. E. M., Porto, D. L., & Caulliraux, H. M. (2014). Supply Chain Integration Researches: An Overview of the Field. *International Journal of Supply Chain Management*, *3*, 12–20.

Spinks, A. W. (2001). Satellite and resort office in Japan. *Transportation, 18*.

Squazzoni, F. (2009). *Epistemological Aspects of Computer Simulation in the Social Sciences*. Springer. doi:10.1007/978-3-642-01109-2

Sterman, J. D. (2000). *Business Dynamics: Systems Thinking and Modeling for a Complex World*. McGraw-Hill/Irwin.

Sterman, J. D. (2002). All Models Are Wrong: Reflections on Becoming a Systems Scientist. *System Dynamics Review*, *18*(40).

Storey, J. (2009). *The Routledge companion to SHRM*. Taylor&Francis.

Sundbo, J. (1994). Modulization of service production and a thesis of convergence between service and manufacturing organizations. *Scandinavian Journal of Management, 10*(3), 245-266.

Sushil. (1993). *System Dynamics: A Practical Approach For Managerial Problems*. Wiley Eastern Ltd.

Sushil. (1993). *Systems Dynamics: A Practical Approach for Managerial Problems*. Wiley Eastern Limited.

Taghezout, N., Adla, A., & Zaraté, P. (2009). A multi-agent framework for group decision support system: Application to a boiler combustion management system (GLZ). *International Journal of Software Engineering and Its Applications*, *3*, 2, 9–20.

Tanasescu, D. A. (2014). *E-Communication and E-development, efficient response for customer expectations into an E-economy environment*. ECR.

Tapamoy, D. (2009). *Compensation management*. Excel Books.

Taylor, S. (2014). *Agent-based Modeling and Simulation*. Springer. doi:10.1057/9781137453648

Tether, B. S. (2002). Who co-operates for innovation, and why: An empirical analysis. *Research Policy*, *31*(6), 947–967. doi:10.1016/S0048-7333(01)00172-X

Tieghi, A. L. (2013). *A saúde brasileira tem cura?* Revista Espaço Aberto USP.

Tushman, M. L., & Romanelli, E. (1985). Organizational evolution: A metamorphosis model of convergence and reorientation. *Research in Organizational Behavior*, *7*, 171–222.

Ucbasaran, D., Westhead, P., & Wright, M. (2006). Opportunity identification and pursuit: Does an entrepreneur's human capital matter? *3rd Specialized Research Workshop, Gate2*. Growth Academic Network, University of Amsterdam.

Uzzi, B. (1996). The sources and consequences of embeddedness for the economic performance of organizations: The network effect. *American Sociological Review*, *61*(4), 674–698. doi:10.2307/2096399

Vähätalo, M., & Kallio, T. J. (2015). Organising health services through modularity. *International Journal of Operations & Production Management*, *35*(6), 925–945. doi:10.1108/IJOPM-12-2013-0523

Van de Ven, A., Polley, D., Garud, R., & Venkataraman, S. (1999). *The Innovation Journey*. Oxford University Press.

Van de Ven, A., & Poole, M. S. (2005). Alternative approaches for studying organizational change. *Organization Studies*, *26*(9), 1377–1404. doi:10.1177/0170840605056907

Van Riet, C., & Demeulemeester, E. (2015). Trade-offs in operating room planning for electives and emergencies: A review. *Operations Research for Health Care*, *7*, 52–69. doi:10.1016/j.orhc.2015.05.005

Vassolo, R. S., Anand, J., & Folta, T. B. (2004). Non-additivity in portfolios of exploration activities: A real options-based analysis of equity alliances in biotechnology. *Strategic Management Journal*, *25*(11), 1045–1061. doi:10.1002/smj.414

Vecina Neto, G., & Malik, A. (2011). *Gestão em Saúde*. Rio de Janeiro: Guanabara Koogan.

Vilkamo, T., & Keil, T. (2003). Strategic technology partnering in high velocity environments-lessons from a case study. *Technovation*, *23*(3), 193–204. doi:10.1016/S0166-4972(01)00108-0

von Raesfeld, A. M., Geurts, P., & Jansen, M. (2012). When is a network a nexus for innovation? A study of public nanotechnology R&D projects in the Netherlands. *Industrial Marketing Management*, *41*(5), 752–758. doi:10.1016/j.indmarman.2012.06.009

Wainer, G., & Mosterman, P. (2016). *Discrete-Event Modeling and Simulation: Theory and Applications*. CRC Press.

Wang, T., Guinet, A., Belaidi, A., & Besombes, B. (2009). Modelling and simulation of emergency services with ARIS and Arena. Case study: The emergency department of Saint Joseph and Saint Luc Hospital. *Production Planning and Control*, *20*(6), 484–495. doi:10.1080/09537280902938605

Wang, W. (1997). Technology transfer to and absorption in a developing world—case studies of Ericsson in Malaysia and China. *Sixth International Conference on Management of Technology*, 25–28.

Ward, N., & Shabha, G. (2001). Teleworking: An assessment of socio-psychological factors. Facilities, 19(1-2).

Wilner, N., & Birnberg, J. (1986). Methodological problems in functional fixation research: Criticism and suggestions. *Accounting, Organizations and Society*, *11*(1), 71–80. doi:10.1016/0361-3682(86)90019-X

Wilson, J. T. (1981). Implementation of computer simulation projects in health care. *The Journal of the Operational Research Society*, *32*(9), 825–832. doi:10.1057/jors.1981.161 PMID:10252405

Xiang, W., Yin, J., & Lim, G. (2015). An ant colony optimization approach for solving an operating room surgery scheduling problem. *Computers & Industrial Engineering*, *85*, 335–345. doi:10.1016/j.cie.2015.04.010

Yıldırım, U. Z., Tansel, B. Ç., & Sabuncuoğlu, İ. (2009). A multi-modal discrete-event simulation model for military deployment. *Simulation Modelling Practice and Theory*, *17*(4), 597–611. doi:10.1016/j.simpat.2008.09.016

Yu, W.-D., & Lo, S.-S. (2009). Patent analysis-based fuzzy inference system for technological strategy planning. *Automation in Construction*, *18*(6), 770–776. doi:10.1016/j.autcon.2009.03.003

Zahradníčková, L., & Vacík, E. (2014). Scenarios as a Strong Support for Strategic Planning. *Procedia Engineering*, *69*, 665–669. doi:10.1016/j.proeng.2014.03.040

Zaim, S., Bayyurt, N., Tarim, M., Zaim, H., & Guc, Y. (2013). System dynamics modeling of a knowledge management process: A case study in Turkish Airlines. *Procedia: Social and Behavioral Sciences*, *99*, 545–552. doi:10.1016/j.sbspro.2013.10.524

Zhang, H., Calvo-Amodio, J. R., & Haapala, K. (2013). A conceptual model for assisting sustainable manufacturing through system dynamics. *Journal of Manufacturing Systems*, *32*(4), 543–549. doi:10.1016/j.jmsy.2013.05.007

Zollo, M., & Winter, S. G. (2002). Deliberate learning and the evolution of dynamic capabilities'. *Organization Science*, *13*(3), 339–351. doi:10.1287/orsc.13.3.339.2780

# Index

## N



## O



## P



## R



## S



## T



## U



## V

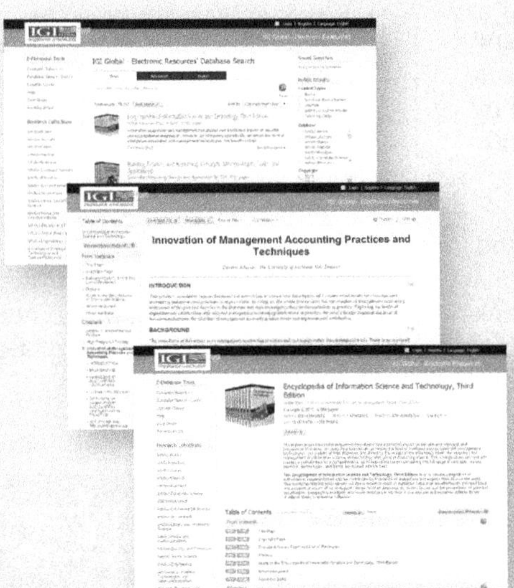

CPSIA information can be obtained
at www.ICGtesting.com
Printed in the USA
LVHW05*1320020718
582494LV00011B/74/P

9 781522 532262